# Functional Nanostructures for Energy Conversion and Advanced Applications

*Edited by Muhammad Rafique, Mahmoud Zendehdel, Muneeb Irshad and Narges Yaghoobi Nia*

Published in London, United Kingdom

Functional Nanostructures for Energy Conversion and Advanced Applications
http://dx.doi.org/10.5772/intechopen.1006220
Edited by Muhammad Rafique, Mahmoud Zendehdel, Muneeb Irshad and Narges Yaghoobi Nia

Contributors
Adnan Khalil, Ahmed Rebey, Alvena Shahid, Ghulam Nabi, Hao Du, Haythem Barrak, Javad Maleki, Kouloud Kourchid, Mahmoud Zendehdel, Maryam Shahrostami, Mojtaba Abdi-Jalebi, Mondher Yahya, Mourad Mbarki, Muhammad Adnan Anwar, Muhammad Asif Nawaz, Muhammad Rafique, Muneeb Irshad, Narges Yaghoobi Nia, Qian Zhang, Ramzi Alaya, Sarmad Masood Shaheen, Siming Huang, Yi Chen, Zhenyu Zhang, Zhu Ma, Zhuo Lv, Zhuowei Du, Zohra Nazir Kayani



First published in London, United Kingdom, 2026 by IntechOpen
IntechOpen is the global imprint of INTECHOPEN LIMITED, registered in England and Wales, registration number: 11086078, 167-169 Great Portland Street, London, W1W 5PF, United Kingdom

For EU product safety concerns: IN TECH d.o.o., Prolaz Marije Krucifikse Kozulić 3, 51000 Rijeka, Croatia, info@intechopen.com or visit our website at intechopen.com.

British Library Cataloguing-in-Publication Data
A catalogue record for this book is available from the British Library

Functional Nanostructures for Energy Conversion and Advanced Applications
Edited by Muhammad Rafique, Mahmoud Zendehdel, Muneeb Irshad and Narges Yaghoobi Nia
p. cm.
Print ISBN 978-1-83635-061-3
Online ISBN 978-1-83635-060-6
eBook (PDF) ISBN 978-1-83635-062-0

# Meet the editors

Dr. Muhammad Rafique is an assistant professor of physics at the University of Sahiwal, Pakistan. He has published more than 100 research articles in well-reputed international journals with high impact factors and has authored many international books with international publishers. His research interests include nanoclusters, nanomaterials, catalysis, energy conversion and storage devices, and related areas. Stanford University, in the United States, has named him a 2% scientist worldwide for his accomplishments.

Dr. Mahmoud Zendehdel is a Professor of Chemistry and Nanotechnology with a Ph.D. in Inorganic Chemistry. He has over 18 years of research and industrial experience in electrochemistry, photochemistry, materials science, and nanotechnology. Dr. Zendehdel has coordinated several high-impact projects in areas such as natural gas desulfurization, methane conversion, and smart energy systems, leading to 14 patents in advanced energy technologies and materials. He is currently the co-founder and CRO/CAO at Iritaly Trading Company S. r. l., where he is managing the "SUNPEROM" HORIZON EIC - Pathfinder Challenges project, "NEXTCCUS" EU project (ERA-NET ACT 2021), and CHOH-D (an industrial Italian project), a senior scientist at the University of Rome Tor Vergata, and CEO of Kimia Solar. His work focuses on innovative solutions in electrochemical systems, perovskite solar modules, and photonic devices.

Dr. Muneeb Irshad is an Assistant Professor of Physics at the University of Engineering and Technology in Lahore, Pakistan. Dr. Muneeb has published over 50 research articles in high-impact international journals. His research interests include nanomaterials, energy conversion and storage devices, catalysis, etc.

Dr. Narges Yaghoobi Nia holds a Ph.D. in Electronics Engineering and is a two-time winner of the prestigious MARIE CURIE Fellowship, awarded in 2014 and 2022 for her work on perovskite technologies for space. She is the founder of the PVSPACE Organisation and PEROSPACE, and the president and co-founder of Iritaly Trading Company S. r. l. She was the Principal Investigator of the P4SPACE project at EPFL in Lausanne and the Aerospace Engineering School of Sapienza University of Rome, with over 13 years of experience in emerging thin-film photovoltaics and electrochemistry. She has published more than 50 scientific papers and holds three patents. Dr. Yaghoobi Nia is also the Editorial Board Manager of *Communications Engineering* in the Nature Portfolio and actively contributes to several EU-funded projects.

# Contents

# Preface

Advances in nanotechnology have enabled the design of functional nanostructures with tailored properties, offering new opportunities for energy conversion, storage, and advanced material applications. This book, *Functional Nanostructures for Energy Conversion and Advanced Applications*, brings together recent developments in self-assembled nanostructures, magnetic nanocomposites, and conducting polymers, integrating both experimental and theoretical perspectives relevant to electrochemical systems.

The first chapter, "Types and Applications of Electrochemical Cells", outlines the principles, classifications and practical uses of electrochemical devices, which form the foundation of modern sustainable energy technologies.

The second and third chapters focus on self-assembly in thin-film photovoltaics, describing how nanoscale structuring enhances light absorption, charge transport, and overall device efficiency. Together, these chapters connect the fundamental mechanisms of self-assembly with applied strategies for improving photovoltaic performance.

The fourth chapter, "Preparation and Magnetic Properties of $Fe_3O_4$@rGO Nanocomposite Investigated by ESR Spectroscopy", highlights the synthesis of hybrid nanostructures and evaluates their electronic and magnetic characteristics using advanced spectroscopic tools, emphasizing their multifunctional role in sensing and energy applications.

The fifth chapter, "Polyaniline and Its Nanocomposites: From Synthesis to Functional Applications", examines conducting polymers, discussing their versatility and practical use as active components in electronic, sensing, and energy storage devices.

The final chapter, "Theoretical Study of Self-Assembled Bi Nanolines on the InAs(001) Surface", provides a theoretical study of Bi nanolines on the InAs(001) surface, offering computational insights into surface nanostructuring and its implications for next-generation nanoelectronic and optoelectronic systems.

Taken together, these contributions offer a comprehensive overview of functional nanostructures, addressing both fundamental understanding and application-driven innovation. The book is intended to serve as a valuable reference for researchers, graduate students, and professionals in materials science, nanotechnology, physics, and energy engineering, while also inspiring further advances in sustainable energy technologies.

We gratefully acknowledge the contributions of the authors, co-authors, reviewers, and editorial staff, whose dedication and expertise have been central to the successful completion of this volume.

**Dr. Muhammad Rafique**
University of Sahiwal,
Pakistan

**Dr. Mahmoud Zendehdel and Dr. Narges Yaghoobi Nia**
Iritaly Trading Company S.r.l.,
Rome, Italy

**Dr. Muneeb Irshad**
University of Engineering and Technology,
Lahore, Pakistan

Chapter 1

# Types and Applications of Electrochemical Cells

*Ghulam Nabi, Muhammad Rafique, Adnan Khalil, Ghulam Nabi, Muneeb Irshad, Muhammad Adnan Anwar, Muhammad Asif Nawaz and Sarmad Masood Shaheen*

## Abstract

This chapter provides a detailed examination of electrochemical cells, starting with the basic concepts of half-cells, reactions at equilibrium, and cell potential. It classifies electrochemical cells as galvanic and electrolytic cells and also includes new technologies like fuel cells and supercapacitors. The mechanism of electron transfer and comparison of cell types is highlighted. The applications portion ventures into classical applications, exploring electrochemical $CO_2$ capture and reduction (ECCR), additive manufacturing, biosynthetic materials, and polymer electrosynthesis. There is a dedicated section on batteries with both conventional systems such as lead-acid and lithium-ion and biofuel and polymer-based derivatives. In addition, the chapter explores how energy density in electrochemical systems is controlled by cell voltage, capacity, and design parameters. Extensive focus is provided to electrodeposition methods throughout metals, alloys, semiconductors, and nanostructures, with ensuing extensive discussions on dimensional control, 2D/3D conformality, and nucleation and growth processes. Parametric factors such as temperature, pH, geometry, and pressure are extensively discussed, providing readers with a synoptic perception of both theory and new applications. The chapter emphasizes the central role of electrochemical systems in meeting current scientific, industrial, and energy challenges.

**Keywords:** supercapacitors, fuel cell, galvanic cell, electrolytic cell, electrochemical sensor, batteries, electrodeposition, nucleation and growth, energy density

## 1. Introduction

An electrochemical cell refers to an apparatus that can either produce electricity through chemical reactions or induce chemical reactions by supplying electricity. A regular 1.50-volt cell for domestic use is one of the best examples of electrochemical cells and is also called single Galvanic Cell (GC). A battery is made up of two or more cells connected either in series or in parallel [1]. In an electrochemical cell, the half-cell reactions, equilibrium conditions, and cell potential are interrelated concepts that determine how the cell produces electrical energy and are being discussed in preceding sections.

### 1.1 Half-cells

An electrochemical cell contains two half-cells, each contributing significantly to the overall reaction. An electrochemical system that interfaces with an electrolyte solution is an electrode, and the unit that interfaces with the electrode is known as a half-cell. The two half-cells may utilize the identical electrolyte solution, or they may utilize distinct electrolyte solutions. Chemical processes that occur within the cell may include chemicals that are found within the electrolyte, electrolytes, or chemicals more distant from the cell, such as hydrogen gas, in the case of fuel cells. In a complete electrochemical cell, one set of half-cell atoms transfers electron to the electrode (this is known as the process of oxidation) and the other set of half-cell atoms involving the electrodes will accept electrons turning to reduced state (this is termed as reduction) [2, 3].

Electrons flow from one half-cell to the other through an external circuit, creating a potential difference. Without any ionic contact, however, the electron flow would eventually be stopped by this charge differential. In order to maintain a steady-state charge distribution between the oxidation and reduction vessels while maintaining their contents otherwise distinct, a salt bridge permits the movement of either positive or negative ions. Separation of solutions can also be achieved using other devices like porous pots and gelled solutions. A porous pot is used in the Bunsen cell (right) [1].

A salt bridge, as shown in **Figure 1**, is commonly described as a filter paper moistened in $KNO_3$, KCl or some other electrolyte. Its function is to make ionic contact between two half-cells that possess distinct electrolytes and prevent the liquids from combining and generate parasitic effects. Such a case is electrolysis of water, where salt bridge is not used and the two half-cells are brought into contact with each other so that the solutions can mix [4].

### 1.2 Equilibrium reaction

Every half-cell has its specific voltage, and distinctive substances used in half-cells result in distinct potential differences. Every reaction occurs at an equilibrium state of

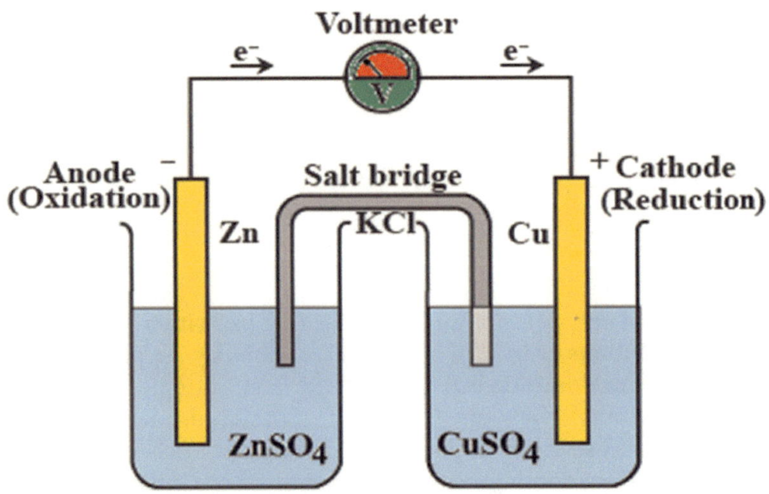


**Figure 1.**
*A Michael Faraday's Daniell type cell is depicted by the schematic.*

several oxidation states of ions: after the cell voltage is not produced. Hence, the half-cell which undergoes oxidation is said to produce more potential as the equilibrium approaches the ion/atom of more positive oxidation states. Raising the potential in the reduction reaction does the same thing, bringing the equilibrium closer to the atom or ion in a more negative oxidation state [1, 5].

### 1.3 Cell potential

Electrode potentials or the voltages of single half-cells can be used to estimate the cell potential. The standard hydrogen electrode (SHE) is assigned 0 volts in order to define these half-cell potentials. A prediction for the observed potential is provided by the voltage differential between electrode potentials. To get the voltage differential, the half-cell reaction must be rewritten in order to produce an equilibrium of the oxidation-reduction [1, 6]. The reactions are as follows.

Reduction half-reaction

$$MnO_4^-(aq) + 8H^+(aq) + 5e^- \rightarrow Mn^{2+}(aq) + 4H_2O(l) \tag{1}$$

Oxidation half-reaction:

$$Sn^{2+}(aq) \rightarrow Sn^{2+}(aq) + 2e^- \tag{2}$$

Balancing the half-reactions

$$5Sn^{2+}(aq) \rightarrow 5Sn^{4+}(aq) + 10e^- \tag{3}$$

The principles in balancing and analyzing redox (reduction-oxidation) reactions are as follows:

1. In order to produce an oxidation reaction or a generally positive cell potential, reverse the reduction process with the least amount of potential [1].

2. To achieve electron balance, it is necessary to multiply half-reactions by integers [1].

Multiplying half-cell reactions by a constant does not change the overall cell potential, as the potential is independent of the reactant coefficient. For most electrochemical cells, the cell potential generally falls within the range of 0 to 6 volts [1, 7].

## 2. Types of electrochemical cells

Electrochemical cells are divided into two types, that is, galvanic and electrolytic cells [8].

### 2.1 Galvanic cells

Chemical energy is immediately transformed into electrical energy in an electrochemical cell. The oxidation and reduction processes that take place inside the cell are what cause this energy conversion. Galvanic cells are electrochemical cells that exhibit spontaneous electrode reduction/oxidation processes. Electrochemical power sources are portable, and scalable. Nowadays, electrochemical cells are used in many different applications, such as powering electric cars and portable electronics. Electrochemical cells are also used to extract metals from aqueous solutions. Metal oxide is created by roasting metal ore in air. It is successively dissolved in the galvanic cell's acidified aqueous solution. The metal can be electrochemically deposited at the cathode by supplying an electrical current to the cell. Metals are extracted electrochemically, including Pb, Zn, Ni, Co, Cd, Cr, Sn, and Mn, etc., as shown in **Figure 2** [9].

Batteries also make use of electrochemical cells. Batteries consist of a number of electrochemical cells Connected in series, parallel, or a mix of both. Batteries can power portable electronics by storing electrical energy from an external source Lead-acid batteries account for the majority of the global battery market, which is worth over $100 billion.

Fuel cells are another type of electrochemical cell of interest. In terms of how they work, fuel cells are comparable to batteries. The location of the reactants is where a battery and a fuel cell differ from one another. In fuel cells, the electrochemical cell is supplied by an external fuel source, whereas in batteries, the fuel is contained within the cell. Over the years, fuel cells have been the subject of much research and used in a various applications, such as powering unmanned underwater vehicles and spacecraft. Specifically, fuel cell vehicles have been powered by hydrogen/oxygen fuel cells. Under open circuit conditions, a single hydrogen/oxygen fuel cell generates about 1 V. To achieve larger voltage values, a stack is created by connecting these fuel cells in series [10, 11].

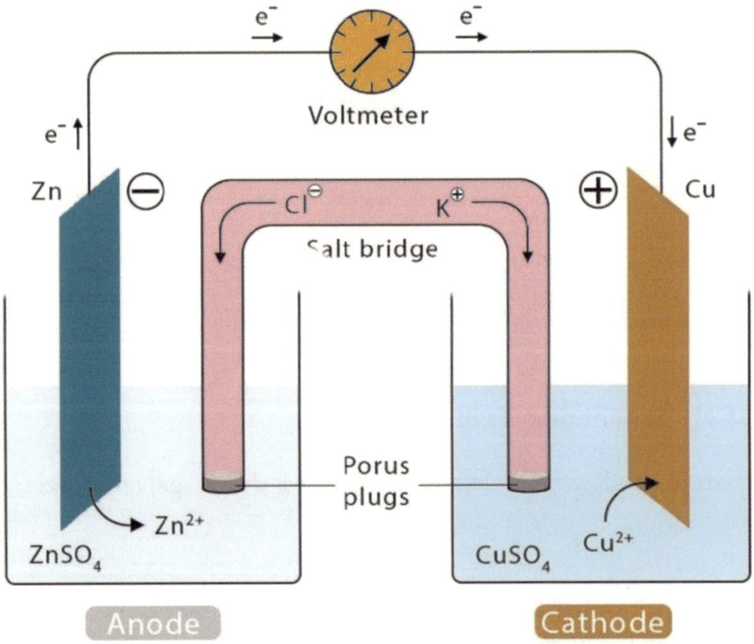


**Figure 2.**
*Schematic of galvanic electrochemical cell.*

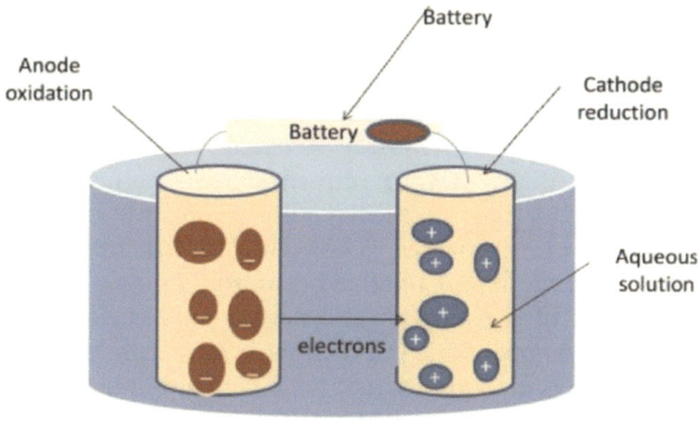


**Figure 3.**
*Schematic of an electrolytic cell [8].*

## 2.2 Electrolytic cell

When electrical energy is delivered to an electrochemical cell, a redox reaction takes place, which is known as an electrolytic cell. Electrolysis, which means to break up (in Greek), is the procedure most frequently employed to break down chemical molecules. The system's chemical energy rises with the addition of electrical energy. Electrolytic cells are often made up of two half-cells, just like galvanic cells.

Electrolytic cells are composed of two electrodes (cathode and anode) and an electrolyte. Dissolved ions are usually found in the solvent solution which serves as the electrolyte and may also consist of molten salts such as sodium sulfate. Faradaic or redox charge-transfer reactions take place when the electrodes are supplied with an external voltage which causes the electrolyte ions to move toward an electrode with a reversal load. Generally, inactive or stable chemical substance in the solution can only be electrolytically dissociated when an external electrical potential of sufficient and correct polarity and magnitude is supplied. A reaction of this kind may cover some other known reactions but may not be possible under the given conditions without the current supply as portrayed in **Figure 3**.

# 3. Applications of electrochemical cells

## 3.1 Electrochemical sensors

The phenomenal progress in semiconductor metal oxide-based sensing technology over the last few decades has drawn the interest of numerous researchers who are working to ensure human safety and security. Air pollution can occur when new substances are released into the Earth's atmosphere, whether from natural sources like wildfires or volcanic activity, or from human-made sources such as factories and vehicle emissions. Air pollution not only damages agricultural food crops and other living things like animals but it also has an adverse effect on human health and causes diseases like cancer, heart disease, lung disease, and mortality that ultimately result in death [12]. There are two types of air pollutants: primary and secondary. Man-made sources often produce primary air pollutants. While secondary pollutants are not immediately released, they include volatile organic compounds (VOCs), nitrogen oxide ($NO_x$), sulfur dioxide ($SO_x$), and carbon monoxide (CO). When primary pollutants interact or react, they are created in the atmosphere. Ground level ozone is the greatest illustration of a secondary air pollutant. Most man-made volatile organic compounds (VOCs) that cause pollution come from Formaldehyde, methyl tert-butyl ether (MTBE), toluene, methylene chloride, benzene, ethyl benzene, and fossil fuels. One of the most widely utilized VOCs in industrial manufacturing and the creation of various chemicals, including plastics, rubber, resins, lubricants, dyes, pharmaceuticals, synthetic textiles, and insecticides, is benzene [13, 14]. Although formaldehyde emissions have an impact on indoor relative humidity [15].

Because of the variety of chemical species they can act as either donor or acceptor charge carriers, single layer graphene (SLG), few-layer graphene (FLG), and graphene oxide (GO) are remarkably sensitive to changes in electrical conductivity, with SLG being the most sensitive to electrical changes in conductivity. Since the atom is a surface atom, each graphene atom is capable of interacting with even a single molecule of a gas or vapor species which increases the effectiveness of a sensor [16–19].

### *3.1.1 Sensors mechanism*

When exposed to a target gas, the thin films' electrical conductivity or resistivity changes, which is how the metal oxide gas sensor operates. As the molecules of the target gas interact with the metal oxide surface and, depending on the target gas, either donate or receive charge carriers, the resistivity or conductivity of metal oxide coatings can change depending on various factors (**Figure 4**).

The electrical resistance of metal oxide thin films can either increase or decrease depending on several factors, including the type of charge carriers present in the material and the nature of the surrounding gas molecules, which can be either

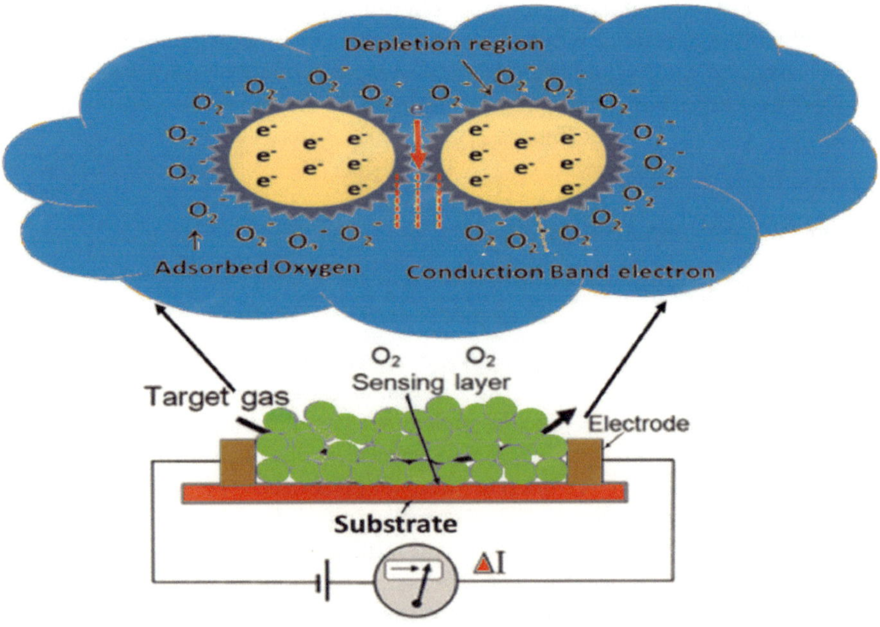


**Figure 4.**
*The metal oxide thin film gas sensor is shown schematically.*

reducing or oxidizing. In n-type semiconductors, electrons are the majority charge carriers, and their conductivity improves in the presence of reducing gases, which donate electrons to the material, increasing the number of charge carriers available for conduction. Conversely, when exposed to an oxidizing gas, electrons are removed from the material, causing a reduction in charge carriers and thus decreasing conductivity. In p-type semiconductors, where positively charged holes serve as the dominant charge carriers, the presence of an oxidizing gas can increase conductivity. This is because oxidizing gases tend to create additional holes by accepting electrons from the material, which enhances the flow of positive charge carriers. Therefore, the interaction with ambient gases plays a significant role in modulating the electrical properties of metal oxide thin films based on their type and the charge carriers involved [20, 21].

### 3.2 Chlor-alkali electrolysis

Given that more than half of all industrial chemical processes employ its outputs, the chlor-alkali process is essential to the chemical industry [22, 23]. However, this industry is known for being highly energy-intensive and for generating pollutants that can negatively affect both the environment and human health [24].

In its most basic form, electrolysis produces two half reactions, the first one known as the HER which occurs at the cathode and the second being the CER where a chloride reaction occurs at the anode while producing sodium hydroxide in the brine solution. Chlor-alkali electrolysis, more commonly known as brine electrolysis, has been using three separate membranes since 1888, which include diaphragm cells, mercury cells, and current membrane cells. Such membranes are useful because they allow for the sodium hydroxide solution to be formed in the cathode compartment during the electrolysis process while also separating the two reactions that consume an equal amount of brine. In the mercury cell, the chemical electroreaction (CER) at the anode is linked to the formation of sodium amalgam at the liquid mercury cathode.

$$2Hg + 2Na^{+} + 2e^{-} \rightarrow 2NaHg \tag{4}$$

Later, the sodium amalgam is extracted and transferred to a second cell, where it undergoes an anodic reaction involving the release of Na + ions, while the cathodic reaction involves the hydrogen evolution reaction (HER), leading to the production of NaOH.

$$NaHg \rightarrow Na^{+} + Hg + e^{-} \tag{5}$$

The membrane cell is comparable to the diaphragm cell, but instead of using a porous asbestos mat, it uses a polymer ion exchange membrane. While the electrolysis process in mercury cells consumes more energy because of the higher cell voltage, it frequently delivers a better end product than either the diaphragm or the membrane cells due to the fact that $Cl_2$ and NaOH were manufactured separately [25–28].

#### *3.2.1 Mechanism of chlor-alkali electrolysis*

As mentioned below and shown in **Figure 5**, Step 1 of the electrolysis process is the production of $H_2$ alkalinated with NaOH, and Step 2 is the generation of $Cl_2$ in saturated NaCl solution. Step 1 contains the cathodic reduction of $H_2O$ on the HER

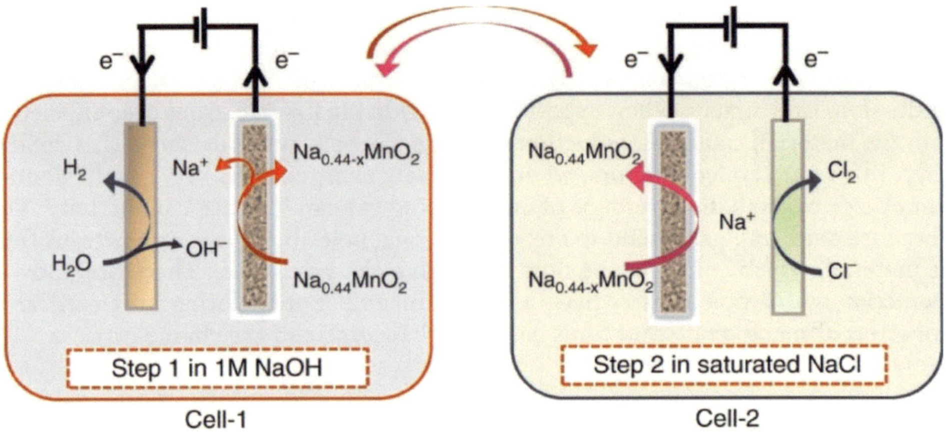


**Figure 5.**
*Illustration of the electrolysis cell's working mechanism [29].*

electrode forming $H_2$ and $OH^-$ (Eq. (2)) and the anodic $Na^+$ de-intercalation from the $Na_{0.44}MnO_2$ electrode (Eq. (1)). In Step 2, the oxidation of molecular $Cl_2$ during the electrolytic synthesis of intercalation compounds in the $Na_{0.44-x}MnO_2$ electrode shut-off the $Na^+$ intercalation into this electrode (Eq. (3)) and also the exponential decay of the anodic $Cl_2$ evolution reaction on the CER electrode (Eq. (4)) at $x > 0.44$ occurs. This is done after the cleaning of Brine the $Na_{0.44-x}MnO_2$ (which is synthing Anodic $Na_{0.44}MnO_2$ during Step 1) which is already removed with brine. $Na_{0.44}MnO_2/Na_{0.44-x}MnO_2$ cell can shift in between cell 1 (for step 1) and cell 2 (for step 2) to repeat Step 1 and Step 2.

Step: 1

$$\text{Anode: } Na_{0.44}MnO_2 \rightarrow Na_{0.44-x}MnO_2 + xNa^+ + xe^- \quad (6)$$

$$\text{Cathode: } 2H_2O + 2e^- \rightarrow H_2 + 2OH^- \quad (7)$$

Step: 2

$$\text{Anode: } 2Cl^- \rightarrow Cl_2 + 2e^- \quad (8)$$

$$\text{Cathode: } Na_{0.44-x}MnO_2 + xNa^+ + xe^- \rightarrow Na_{0.44}MnO_2 \quad (9)$$

This method produces a chlor-alkali electrolysis process that has a number of significant benefits.

- First, compared to diaphragm cells, mercury cells, and even the existing membrane cells, this technology is cleaner due to the ecologically favorable nature of $Na_{0.44}MnO_2$.

- Secondly, this design does not require a membrane to generate $H_2$ (+NaOH) and $Cl_2$ independently.

Therefore, with the separate production of $Cl_2$ and $H_2$ no product gases mix at all current densities, gas handling is simplified greatly and thus the operational flexibility of chlor-alkali electrolysis cells is largely increased [30–34].

### 3.3 Electrochemical $CO_2$ capture and reduction (ECCR)

Electrochemical $CO_2$ capture and reduction (ECCR) is a novel technology that operates based on electrochemical cells to capture carbon dioxide ($CO_2$) from the air or industrial flue gases and convert it into useful fuels and chemicals like carbon monoxide (CO), formic acid (HCOOH), methane ($CH_4$), ethylene ($C_2H_4$), and ethanol ($C_2H_5OH$). This process provides a promising route toward carbon neutrality through greenhouse gas mitigation and production of sustainable energy carriers and industrial feedstocks. ECCR systems take advantage of electrocatalysts, for example, copper, silver, or metal-organic frameworks, in order to enable $CO_2$ reduction at the cathode and water or oxygen evolution reactions at the anode. Running at ambient temperatures with renewable power, ECCR operates with efficiencies of 20–50% and Faradaic efficiencies of up to 90% for certain products. Scalability, compatibility with renewable power, and $CO_2$ valorization make it vital for use in energy storage, chemical production, and climate change mitigation applications, though catalyst durability and selectivity are among the challenges motivating current research [35].

#### *3.3.1 Electrochemical reaction mechanisms*

Electrochemical $CO_2$ capture and reduction entails capturing $CO_2$ (air, flue gas, or dissolved sources) and electrochemically reducing it at the cathode to create desired products, with an oxidation reaction on the anode to balance the circuit. The process is as follows: on the cathode, $CO_2$ molecules are adsorbed on the surface of the electrocatalyst where they react with protons ($H^+$) and electrons from the external circuit to produce reduced products such as CO, formic acid, or hydrocarbons. The reaction path varies depending on the catalyst, applied potential, and electrolyte (e.g., aqueous $KHCO_3$ or ionic liquids). Water or hydroxide ions at the anode are oxidized, usually generating oxygen ($O_2$) and protons or water, in balancing the electrochemical cell. Mediators or membranes (such as anion exchange membranes) can enhance $CO_2$ capture by binding selectively to $CO_2$, increasing efficiency. The entire process is powered by an applied voltage (usually 2–3 V) and is controlled by product selectivity based on catalyst properties and operating conditions (**Figure 6**) [37, 38].

Cathode reaction (reduction of $CO_2$ to CO):

$$CO_2 + 2H^+ + 2e^- \rightarrow CO + H_2O \tag{10}$$

Anode reaction (water oxidation):

$$2H_2O \rightarrow O_2 + 4H^+ + 4e^- \tag{11}$$

Overall reaction:

$$2CO_2 + 2H_2O \rightarrow 2CO + O_2 + 2H_2O \tag{12}$$

Electrochemical $CO_2$ capture and reduction is used in various industries to mitigate carbon emissions and create sustainable products. In the chemical industry, ECCR generates feedstocks such as carbon monoxide and formic acid for use in the synthesis of polymers, pharmaceuticals, and industrial chemicals, with formic acid yields of ~90% Faradaic efficiency using tin-based catalysts. In energy storage, $CO_2$ conversion to fuels such as methane or ethanol allows for the storage of renewable

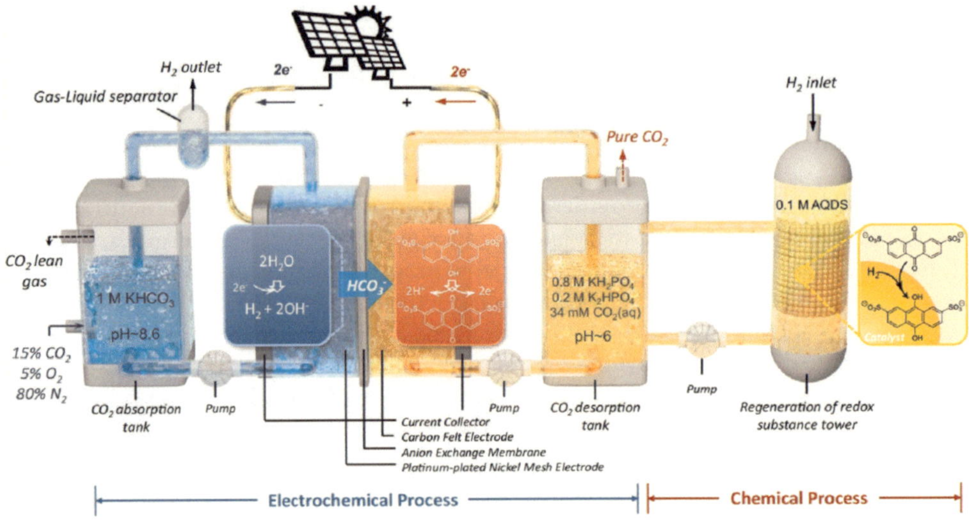


**Figure 6.**
*Schematic diagram of the decoupled redox electrochemical $CO_2$ capture system [36].*

energy (e.g., solar or wind power) in chemical form, enabling carbon-neutral fuel cycles. Environmental uses involve direct air capture (DAC) systems, in which ECCR removes low-concentration $CO_2$ ($\sim$420 ppm) from air, lowering greenhouse gas emissions. Power plant or cement factory industrial emissions are also targeted, with ECCR systems incorporated to capture flue gas $CO_2$ and process it into useful products for economic feasibility [39].

## 3.4 Additive manufacturing

Additive manufacturing (AM) by electrochemical methods commonly termed electrochemical 3D printing is a sophisticated manufacturing technique based on localized electrodeposition for the production of geometrically well-controlled micro- and nanoscale structures from metals, alloys, and functional oxides. In contrast to conventional 3D printing which is based on material extrusion or laser sintering, electrochemical AM exploits redox reactions for material deposition voxel-by-voxel or layer-by-layer, providing superior control of geometry, composition and microstructure. This method provides the capability of producing intricate 3D architectures with feature sizes as low as nanometers, rendering it suitable for high-precision applications, including MEMS, biomedical implants, and printed electronics [40].

### *3.4.1 Electrochemical reaction mechanisms*

Electrochemical 3D printing is based on regulated electrodeposition, wherein metal ions or precursors in an electrolyte are reduced at a cathode to build solid structures with the completion of the circuit by an oxidation reaction at the anode. On the cathode, localized voltage or current application forces reduction of metal ions or precursors and deposits material with high spatial control, usually with the guidance of a scanning probe, nozzle, or template. The anode usually gets oxidized, e.g., water splitting, to balance the electrochemical cell. Methods such as electrohydrodynamic redox printing (EHD-RP) or scanning electrochemical microscopy (SECM) utilize micro/nozzles or probes to localize deposition at desired positions, with submicron resolution. The operation is controlled by factors such as applied potential, electrolyte solution (e.g., $CuSO_4$ for copper), and probe translation, which determine deposition rate and feature dimensions. As an instance, during copper printing, the cathode reaction metallizes, while the anode produces oxygen, allowing continuous material deposition [41].

Cathode reaction (deposition of copper):

$$Cu^{2+} + 2e^{-} \rightarrow Cu\ (s) \tag{13}$$

Anode reaction (oxidation of water):

$$2H_2O \rightarrow O_2 + 4H^{+} + 4e^{-} \tag{14}$$

Overall reaction:

$$2Cu^{2+} + 2H_2O \rightarrow 2Cu\ (s) + O_2 + 4H^{+} \tag{15}$$

Electrochemical 3D printing is revolutionizing numerous high-precision applications owing to its capability to print complex, functional micro- and nanostructures. In printed electronics, it makes it possible to print conductive traces, interconnects, and antennas at submicron resolution, which allows flexible and wearable devices such as sensors and RFID tags. Biomedical applications take advantage of the fabrication of micro-devices [22].

### 3.5 Electrochemical synthesis of biomaterials

Electrochemical cells play a growing role in biomaterial synthesis for medical implantation, tissue engineering, and bioactive coating, providing controlled, scalable, and biocompatible fabrication processes. Working in mild conditions (e.g., physiological pH and room temperature), these processes allow controlled deposition without degrading sensitive biological substrates. Electrochemical synthesis allows the production of bio-ceramics such as hydroxyapatite ($Ca_{10}(PO_4)_6(OH)_2$), antibacterial chitosan films, and conductive polymer films such as polypyrrole (PPy) and poly(3,4-ethylenedioxythiophene) (PEDOT). These materials promote osseointegration, minimize infection risk, and facilitate tissue regeneration, with adhesion strengths of 20–30 MPa in the case of hydroxyapatate on implants. Low-temperature operation, compatibility with intricate 3D geometries, and the capability of tailoring coating properties render the process suitable for biomedical applications. Coating durability, bioactivity, and functionalization are being continuously improved to meet enhanced demands on medical devices and regenerative medicine [23].

#### *3.5.1 Electrochemical reaction mechanisms*

Electrochemical synthesis of biomaterials involves redox processes and ion migration in water or biocompatible electrolytes and allows controlled deposition or polymerization on electrode surfaces. Reduction reactions at the cathode raise local pH to induce precipitation of bio-ceramics such as hydroxyapatite or trigger gelation of biopolymers such as chitosan. Oxidation reactions at the anode, such as splitting of water or polymerization of monomers, counterbalance the cell. For conductive polymers, electropolymerization oxidizes the monomer to a radical intermediate, resulting in chain growth and film formation. These reactions are controlled with current or potential to exactly the same extent, with electrolytes (e.g., $CaCl_2$, $Na_2HPO_4$, or solutions of monomers) being designed to provide biocompatibility and uniform deposition.

Cathode reaction (hydroxyapatite deposition):

$$2H_2O + 2e^- \rightarrow H_2 + 2OH^- \tag{16}$$

Formation reaction (hydroxyapatite):

$$10Ca^{2+} + 6PO_4{}^{3-} + 2OH^- \rightarrow Ca_{10}(PO_4)_6(OH)_2\downarrow \tag{17}$$

Anode reaction (water oxidation):

$$2H_2O \rightarrow O_2 + 4H^+ + 4e^- \tag{18}$$

Overall reaction (hydroxyapatite):

$$10Ca^{2+} + 6PO_4{}^{3-} + 2H_2O \rightarrow Ca_{10}(PO_4)_6(OH)_2 + 4H^+ + O_2 \tag{19}$$

Electrochemical synthesis of biomaterials is widely used in biomedical engineering to produce functional coatings and structures for medical and regenerative uses. Bone implants that have been coated with hydroxyapatite replicate the composition of natural bone, promoting osseointegration and finding success in over 95% of orthopedic implants such as hip replacements and dental screws. Antibacterial surfaces, created through chitosan deposition or silver nanoparticle integration, minimize infection risks by up to 99% in implants, essential for long-term biocompatibility. Tissue engineering scaffolds, created using electrochemically deposited hydroxyapatite, chitosan, or conductive polymers like PPy, provide porous 3D frameworks that support cell proliferation and tissue regeneration for bone, cartilage, and neural repair. Conductive polymer coatings, such as PEDOT, are used in neural implants and biosensors, leveraging their electrical properties to stimulate tissue or monitor biological signals [42].

### 3.6 Electrosynthesis of polymeric materials

Electrosynthesis of polymeric materials encompasses electrochemical routes to polymerize monomers, including aniline, pyrrole, and thiophene, to give conductive polymers, including polyaniline (PANI), polypyrrole (PPy), and poly(3,4-ethylenedioxythiophene) (PEDOT). Such polymers are grown in situ on electrode surfaces via oxidative polymerization, allowing direct integration into devices with fine control over film thickness, morphology, and conductivity (usually 10–1000 S/cm). Conductive polymers bridge metal-like electrical properties and plastic-like flexibility and processability, making them most suitable for flexible electronics, energy storage, and biomedical devices. Carrying out operations at mild conditions (room temperature and aqueous or organic electrolytes), electrosynthesis has some benefits over chemical polymerization, such as scalability, green chemistry, and substrates compatibility. Current research is aimed at increasing polymer stability, conductivity, and functionalization to expand their applications in novel technologies.

### *3.6.1 Electrochemical reaction mechanisms*

Conductive polymers are prepared by electrochemical oxidation of monomers via oxidative polymerization, where the applied potential oxidizes monomers at the anode to give radical cations that start chain growth. Oxidative polymerization starts with the monomer (e.g., aniline and pyrrole) losing an electron at the anode to give a radical cation. These reactive intermediates dimerize by coupling, further oxidizing and coupling to initiate polymer chain growth on the electrode surface. The reaction is commonly performed in an electrolyte (such as $H_2SO_4$ or NaCl for aqueous solutions, $LiClO_4$ in acetonitrile for organic solution) with a counter electrode that enables reduction (such as water reduction) to balance the cell. Polymer films directly deposit on the anode, with the thickness regulated by current, potential, or deposition time. Conductivity and stability are improved by doping with anions or cations during polymerization [43, 44].

Anode reaction (synthesis):

$$\text{Aniline} \rightarrow \text{PANI} + 2H^+ + 2e^- \quad (20)$$

Cathode reaction (reduction):

$$2H_2O + 2e^- \rightarrow H_2 + 2OH^- \quad (21)$$

Conductive polymers electrosynthesized are crucial in a range of sophisticated applications because of their specific electrical and mechanical behavior. As high-surface-area electrodes in flexible supercapacitors, polymers PPy and PEDOT provide specific capacitances of 100–400 F/g and allow for lightweight, flexible energy storage for wearable electronics. Artificial muscles make use of electro-active polymers like PANI or PPy that change their size when they are subjected to applied potentials, allowing for soft robotics and prosthetic devices up to 10% actuation strains. Organic electrochemical transistors (OECTs) make use of PEDOT-based films for bioelectronics applications, e.g., neural interfaces and biosensors, to take advantage of their high transconductance (up to 10 mS) in sensing physiological environmental signals [45, 46].

## 3.7 Batteries

An ion-conducting electrolyte is in contact with two (either identical or different) electron-conducting electrodes in batteries, which are a type of galvanic cell. A battery is a device designed to convert chemical reaction energy between two solid reactants into electrical energy through electrochemical means. Batteries fall into two major categories: primary and secondary.

- Primary batteries are known for their excellent energy storage capabilities and high energy density, but they cannot be recharged. These batteries are designed for single-use, and once the energy is depleted, they must be replaced. Examples of primary batteries include lithium-carbon monofluoride, lithium-thionyl chloride, and lithium-manganese dioxide, each offering different performance characteristics suited for various applications.

- Secondary batteries are more sustainable than primary batteries because they can be regenerated and charge and discharge multiple times. Common examples of

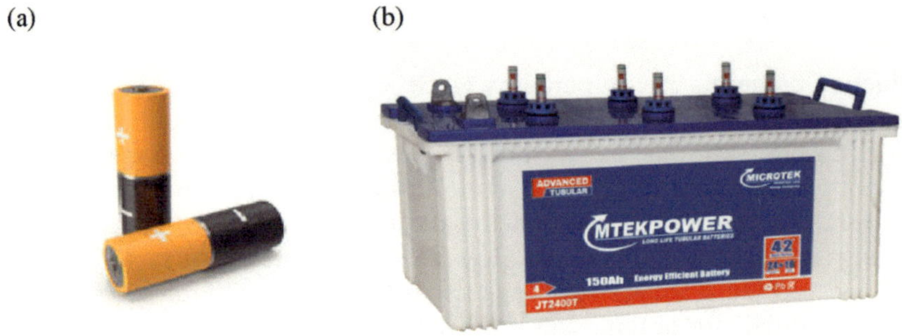


**Figure 7.**
*Batteries. (a) Primary battery. (b) Secondary battery.*

secondary batteries in use today include coin-type lithium secondary batteries, nickel-metal hydride (NiMH) batteries, nickel-cadmium (NiCd) batteries, lead-acid batteries, and lithium-ion batteries. Each type has specific benefits and uses, with lithium-ion batteries being widely used in electric vehicles and portable electronics because of their high energy density and long lifespan (**Figure 7**) [47, 48].

## 4. Electrochemical cell energy density

Energy density, which is one of the most important figures of merit for electrochemical energy storage devices such as batteries and supercapacitors, is a measure of the energy stored per unit of mass (Wh/kg) or volume (Wh/L). High energy density allows for small, light devices to provide long-term power output, and thus it is a requirement in applications such as portable devices, electric vehicles, aerospace, and grid-scale energy storage. Energy density is affected by voltage of the cell, capacity, material characteristics, and structural configuration, with trade-offs against power density commonly dictating the optimization process. High energy density can only be achieved through the judicious choice of electrode materials, electrolytes, and cell designs balanced with efficiency, safety, and cycle life [49].

### 4.1 Cell voltage (V)

The open circuit voltage (OCV) of an electrochemical cell is given by the difference between the cathode and anode standard electrode potentials, as per the Nernst equation:

$$E = E^{O} - \left(\frac{RT}{nF}\right) \ln Q \tag{22}$$

Where E is cell voltage, $E^0$ is standard cell potential, R is the gas constant (8.314 J/mol·K), T is temperature (K), n is the electrons transferred, F is Faraday's constant (96,485 C/mol), and Q is the reaction quotient of ion concentrations. Increased cell voltages directly increase energy density since energy is determined as:

$$\text{Energy (Wh)} = \text{Voltage (V)} \times \text{Capacity (Ah)} \tag{23}$$

For example:

i. A lithium-ion battery with lithium cobalt oxide ($LiCoO_2$, $E^0 \approx 3.9$ V vs. $Li/Li^+$) and graphite ($E^0 \approx 0.2$ V vs. $Li/Li^+$) produces an OCV of ~3.7 V, which leads to high energy density (~150–250 Wh/kg).

ii. A lead-acid battery, using $PbO_2$ (cathode) and Pb (anode), works at ~2.0 V, which gives lower energy density (~30–50 Wh/kg).

Voltage stability depends on electrolyte selection, which needs to provide a large electrochemical stability window to avoid decomposition at high potential [50, 51].

### 4.2 Capacity (Ah) or specific capacity (mAh/g)

Capacity is the charge stored and discharged by a cell, in ampere-hours (Ah) or specific capacity (mAh/g for mass, $mAh/cm^3$ for volume). It is the function of the number of electrons transferred per redox reaction (n), the molar mass of the active material (M), and the quantity of active material. The theoretical specific capacity is expressed as follows:

$$\text{theoretical capacity} = \frac{n \times F}{3.6 \times M} \text{ (in mAh/g)} \quad (24)$$

Where n is the number of electrons, F is Faraday's constant (96,485 C/mol), and M is the molar mass (g/mol). Materials with high n and low M yield higher capacities [52]. For example:

i. Lithium metal: n = 1, M = 6.94 g/mol, theoretical capacity ≈ 3860 mAh/g.

ii. Graphite: n = 1 per 6 C atoms, M = 72 g/mol ($C_6$), theoretical capacity ≈ 372 mAh/g.

### 4.3 Design and structural factors

Aside from chemistry, physical and structural design in electrochemical cells also has a significant effect on energy density:

Electrode thickness and porosity: Thick electrodes enhance active material loading, increasing capacity but potentially impeding ion transport and decreasing efficiency when too thick.

Electrolyte choice: Electrolytes with broad voltage stability windows (e.g., solid electrolytes, ionic liquids) enable high-voltage cathodes, increasing energy density.

Cell Structure: Pouch cells have highest active material volume, thus achieving higher volumetric energy density (~500–700 Wh/L in Li-ion), while cylindrical cells (e.g., 18,650) provide strong packaging but reduced volumetric efficiency. Solid-state structures reduce inactive components, increasing energy density further.

## 5. Classification of electrochemical batteries

Batteries may be categorized according to their chemistry, structure, ability to be recharged, and mode of operation.

## 5.1 Lithium-ion batteries (Li-ion)

Lithium-ion (Li-ion) batteries are light, high-energy-density rechargeable systems essential for supplying power to portable electronics, electric vehicles (EVs), and energy storage at grid scale. They have gravimetric energy density 150–300 Wh/kg, volumetric energy density 400–800 Wh/L, long cycle life (500–2000 cycles), and low self-discharge rate (~2–5% per month). These batteries work on a principle of reversibly transporting lithium ions between an anode (commonly graphite) and a cathode (like lithium cobalt oxide) by means of an electrolyte, for example, $LiPF_6$ dissolved in organic solvents like ethylene carbonate or dimethyl carbonate. Viable in a range of forms (cylindrical, prismatic, and pouch), Li-ion batteries are flexible, with continued innovation in electrode materials, electrolytes, and cell structures advancing performance, safety, and sustainability in solving challenges such as cost, resource limitation, and environmental concerns [53].

### 5.1.1 Electrochemical reaction mechanisms

The Li-ion battery operation is based on reversible electrochemical reactions between the anode and cathode, where lithium ions travel in the electrolyte and electrons in an external circuit. The reaction process occurs through the following steps:

*Discharge process:*

- At the anode, lithium ions are removed from the graphite lattice, surrendering electrons to the external circuit.
- At the cathode, lithium ions intercalate into the cathode material, gaining electrons from the circuit.
- The electrolyte allows lithium-ion transfer between the electrodes, while a separator ensures that electrons do not flow directly, short-circuiting prevented.

*Charge process:*

- On charging, the reverse process occurs: lithium ions are removed from the cathode and reintercalated into the anode, electrons returning through the circuit [54, 55].

An applied voltage of 3.6–4.2 V depending on the cathode material initiates the reactions (**Figure 8**).

Anode reaction:

$$LiC_6 \rightleftharpoons Li^+ + e^- + C_6 \tag{25}$$

Cathode reaction:

$$LiCoO_2 + xLi^+ + xe^- \rightleftharpoons Li_{1-x}CoO_2 \tag{26}$$

Overall reaction:

$$C_6 + LiCoO_2 \rightleftharpoons Li_{1-x}CoO_2 + Li_xC_6 \tag{27}$$

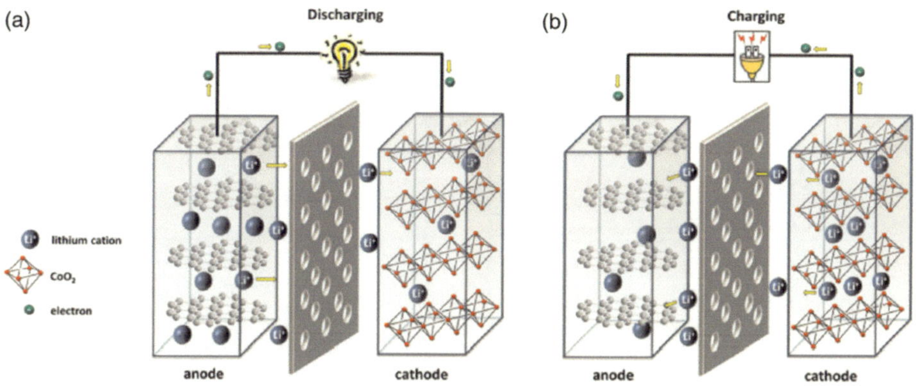


**Figure 8.**
*(a) Discharging and (b) charging of lithium-ion battery [56].*

Li-ion batteries are central to numerous applications owing to their high energy density, efficiency, and versatility. In portable electronics, they supply smartphones, laptops, tablets, and wearables with compact energy for prolonged use (e.g., a 4000 mAh smartphone battery offers 10–20 Wh). Electric cars use Li-ion batteries like the nickel manganese cobalt (NMC) cathode ones (~260 Wh/kg) to provide driving ranges of 200–500 km while still being efficient and powerful. High discharge rates (up to 20C) and fast recharging aid power tools like drills and saws [57].

## 5.2 Lead acid battery

Lead-acid batteries remain the most widely used type of storage battery due to their reliable performance, high dependability, and cost-effectiveness. While certain battery technologies boast a cycle life exceeding a thousand cycles, lead-acid batteries typically offer only a few hundred charge-discharge cycles. The invention of the first practical lead-acid cell dates back to 1859, credited to French scientist Gaston Planté. His design consisted of two lead plates, separated by a strip of cloth, and immersed in a jar containing sulfuric acid. During the initial charging cycle, electrochemical reactions formed a layer of lead dioxide on the surface. Later advancements introduced electrodes made by applying a mixture of sulfuric acid and lead oxides onto lead-antimony alloy grids, commonly referred to as plates in lead-acid batteries [58, 59].

### *5.2.1 Electrochemical reaction mechanisms in the electrodes*

In general, the electrochemical reaction at the negative electrode is written as

$$Pb + HSO_4^- = PbSO_4 + H^+ + 2e^- \quad (28)$$

And that at the positive electrode as

$$PbO_2 + 3H^+ + HSO_4^- + 2e^- = PbSO_4 + 2H_2O \quad (29)$$

It is evident that $H^+$ ions play a role in all of these processes. These ions move through the $PbSO_4$ solid electrolyte surface layers, entering and exiting the electrodes as protons. The ions' movement in and out of the local aqueous environment causes

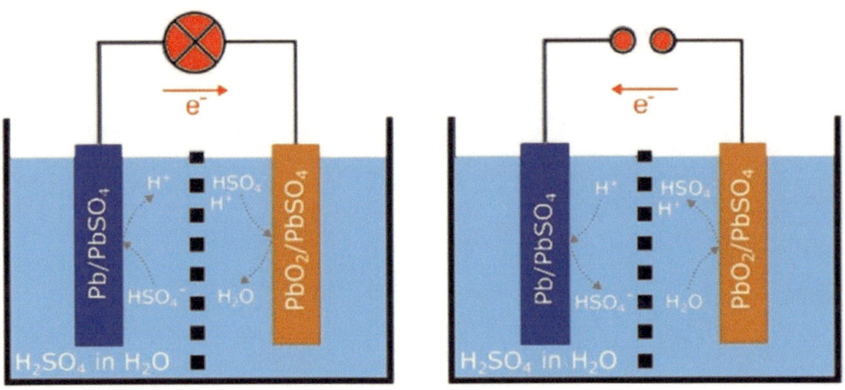


**Figure 9.**
*Discharging and recharging of lead acid battery.*

pH variations, which affect the solubility of $PbSO_4$ in the negative electrode and that of $PbO_2$ and $PbSO_4$ in the positive electrode. As the cell reaction occurs, these solid phases precipitate and/or dissolve. This process is commonly referred to as a dissolution precipitation mechanism (**Figure 9**).

As a result, all the things considered reaction in this type of battery involves the transport of proton ions through a dense $PbSO_4$ layer in the solid electrolyte. The solubility of $PbSO_4$ and $PbO_2$ close to the electrolyte is impacted by this transport's local pH change [60, 61].

### 5.3 Enzymatic biofuel cells (EBFCs)

Enzymatic biofuel cells (EBFCs) are new electrochemical devices that harness enzymes to transform chemical energy in biofuels, for example, glucose or alcohols, into electrical energy via biocatalytic reactions. EBFCs offer a sustainable, biocompatible energy source, running at benign conditions (e.g., physiological pH and temperature) with renewable feedstocks. Using enzymes such as glucose oxidase or laccase as catalysts, EBFCs realize high reaction specificity and efficiency relative to metal catalysts. With power densities as low as 10 $\mu W/cm^2$ to 1 $mW/cm^2$, they are well-suited for low-power devices such as implantable medical devices and biosensors [62].

#### *5.3.1 Electrochemical reactions*

In the anode, the biofuel is most commonly enzymes like glucose oxidase (GOx) or glucose dehydrogenase (GDH) oxidize glucose. The enzyme mediates the transfer of electrons from the biofuel to the electrode, usually via a mediator like ferrocene or quinones or through direct electron transfer (DET) when the enzyme is immobilized near the electrode surface. This operation liberates protons ($H^+$) and electrons, and the electrons travel through the external circuit to create current, leaving gluconolactone as a waste product. Oxygen ($O_2$) or some other oxidant is reduced at the cathode by enzymes such as laccase or bilirubin oxidase (BOD), which drive the reaction. Electrons from the external circuit react with protons and oxygen to create water, which completes the electrochemical cycle. Effective electron transfer is facilitated by mediators or DET, and the cathode is made to function effectively under oxygen-rich conditions, that is, physiological fluids (**Figure 10**) [64, 65].

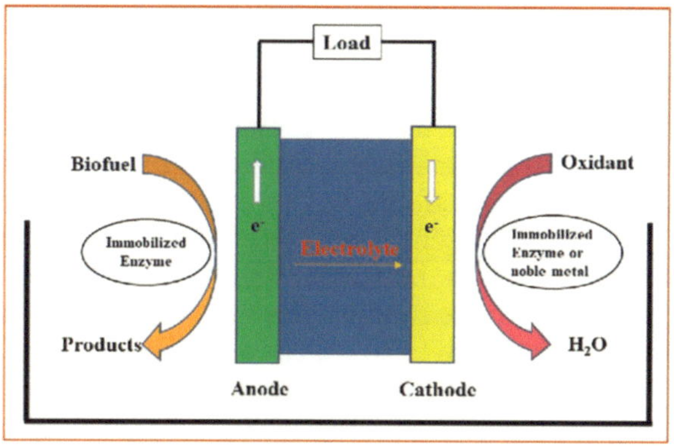


**Figure 10.**
*Principles and configurations of enzymatic biofuel cells [63].*

Anode reaction:

$$\text{Glucose} \rightarrow \text{gluconolactone} + 2H^+ + 2e^- \ (\text{glucose oxidase catalyzes}) \tag{30}$$

Cathode reaction:

$$O_2 + 4H^+ + 4e^- \rightarrow 2H_2O \ (\text{laccase or bilirubin oxidase catalyzes}) \tag{31}$$

Overall reaction:

$$2\text{Glucose} + O_2 \rightarrow 2\text{gluconolactone} + 2H_2O \tag{32}$$

Enzymatic biofuel cells are ideally suited to low-power, biocompatible devices, especially in biomedical and sensor technologies. They drive implantable biosensors for the monitoring of physiological parameters, including blood glucose levels in diabetic patients, using glucose in body fluids to power electricity (e.g., 10–100 μW/cm$^2$) [66].

## 5.4 Polymer-based battery

Polymer battery is a new generation electrochemical energy storage device which applies solid or gel polymer electrolytes instead of conventional liquid electrolytes, with improved safety, flexibility, and light weight. The battery can use conductive polymers, for example, polyaniline (PANI), polypyrrole (PPy) as electrode materials, or incorporate polymer matrices as electrolytes, as in the case of lithium-polymer (Li-Po) batteries. With energy densities of 100–250 Wh/kg and thin-film processability, they are best suited for contemporary electronics, wearables, and biomedicine. They can be integrated into flexible devices due to their ability to bend and are safer compared to liquid-electrolyte systems in terms of leakage and flammability. Continuous studies of redox-active polymer electrodes and solid-state electrolytes will further enhance energy density, cycle life, and scalability and thereby render polymer-based batteries a viable next-generation energy storage solution [56].

### 5.4.1 Electrochemical reaction mechanisms

Polymer-based batteries work through reversible electrochemical reactions, differing in mechanism depending on whether they have polymer electrolytes or redox-active polymer electrodes. In polymer electrolyte-type batteries, like lithium-polymer (Li-Po) batteries, the reactions are similar to lithium-ion batteries but use a gel or solid polymer electrolyte (like polyethylene oxide with $LiPF_6$) to provide ion conductivity. Lithium ions are stripped from graphite at the anode during discharge, releasing electrons to the outside circuit, and lithium ions intercalate into the cathode material, accepting electrons, at the cathode. In conductive polymer-based batteries, electrodes are conductive polymers such as PANI or PPy, which exhibit reversible redox reactions by donating or accepting electrons, usually accompanied by proton or anion doping/undoping. The reactions take place at lower voltages (1–3 V) but provide flexibility and environmental stability (**Figure 11**) [68].

Anode reaction (Li-Po):

$$LiC_6 \rightarrow Li^+ + C_6 + e^- \tag{33}$$

Cathode reaction (Li-Po):

$$Li_{1-x}CoO_2 + xLi^+ + xe^- \rightarrow LiCoO_2 \tag{34}$$

Overall reaction (Li-Po):

$$C_6 + LiCoO_2 \rightarrow Li_{1-x}CoO_2 + Li_xC_6 \tag{35}$$

Polymer-based batteries are extensively applied in flexible, lightweight, and safe applications. In wearables, they drive smartwatches, fitness trackers, and electronic textiles with their bendable character and energy density (100–200 Wh/kg) to adapt to body curves. Flexible components like bendable screens and medical sensors are

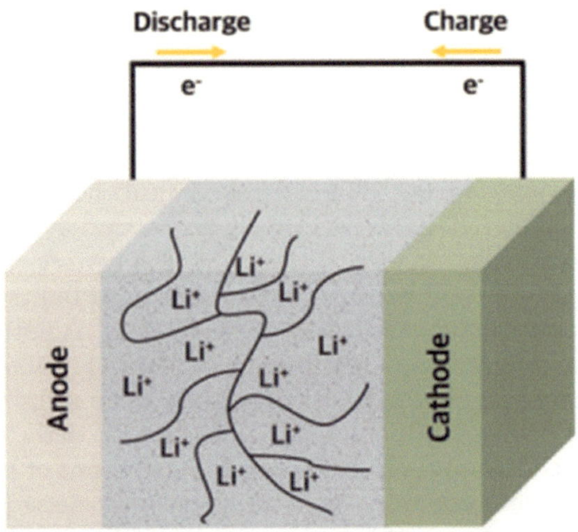


**Figure 11.**
*Schematic diagram of lithium batteries using gel polymer electrolyte [67].*

aided by their thin-film compatibility and mechanical stress resistance, which allows them to be maintained under performance [69].

## 6. Electrodeposition techniques

Electrodeposition the foundation of electrochemical manufacturing is a versatile and scalable method for the deposition of thin films, coatings, and nanostructures with applications in electronics, energy and biomedicine industries. Through application of an electric potential or current metal or compound ions from an electrolyte solution are reduced and deposited on a conductive substrate to produce uniform layers with thickness, composition, and morphology at will. Its affordability capacity to coat intricate geometries and compatibility with a wide range of materials metals, alloys, semiconductors and composites makes it invaluable.

### 6.1 Electrodeposition of metals

Electrodeposition of metals is an umbrella technique in material science and industrial processes, famous for producing uniform, adherent, and high-purity metallic coatings on a wide range of substrates. The electrochemical process is prized for its accuracy, scalability, and versatility to facilitate uses in electronics, corrosion protection, decorative plating, and advanced manufacturing. Through the application of an electric current or potential, metal cations in an electrolyte solution are reduced and deposited onto a cathodic surface, creating a cohesive metallic film with customized properties [70].

#### *6.1.1 Electrochemical reaction mechanism*

The essence of metal electrodeposition is based on a redox reaction, in which metal ions in the electrolyte pick up electrons at the cathode to create a solid metallic film:

$$M^{n+} + ne^- \rightarrow M(s) \tag{36}$$

For instance, in copper electrodeposition, the reaction is

$$Cu^{2+} + 2e^- \rightarrow Cu(s) \tag{37}$$

This reaction takes place in an electrochemical cell consisting of two electrodes a cathode and an anode submerged in an aqueous solution of electrolyte, commonly metal salts (such as $CuSO_4$ for copper plating). The substrate where the metal is plated is the cathode, and the anode may be inert (e.g., platinum, which is insoluble) or soluble (e.g., a copper anode that supplies metal ions to the electrolyte) [71].

### 6.2 Electrodeposition of alloys

Alloy electrodeposition is a complex electrochemical method consisting of or sequential metal ion deposition of two or more ions from an electrolyte solution onto a conductive substrate to produce a solid solution, intermetallic compound, or composite layer. This method is essential for the production of coatings and films with designed properties, including improved corrosion resistance, magnetism, or mechanical properties that can outperform single-metal coatings. Complexity in alloy electrodeposition stems from the interplay between electrochemical characteristics, ion interactions, and process parameters and thus is a versatile yet complex technique used for applications in electronics, magnetic devices, and protective coatings.

### *6.2.1 Electrochemical reaction mechanism*

Alloy electrodeposition mostly takes place via co-deposition, where various metal ions are reduced simultaneously at the cathode. Deposition behavior depends on the electrochemical characteristics of the metals, their reduction potentials, and interactions in the electrolyte. Two main mechanisms are as follows:

*Normal Co-deposition:* In this thermodynamically anticipated process, the nobler metal (with greater reduction potential) co-deposits preferentially. For instance, in a Cu-Ni system, copper ($Cu^{2+} + 2e^- \rightarrow Cu$, $E^0 \approx +0.34$ V) will co-deposit more easily than nickel ($Ni^{2+} + 2e^- \rightarrow Ni$, $E^0 \approx -0.25$ V) under standard conditions, consistent with their relative nobility.

*Anomalous Co-deposition:* In a process opposite to thermodynamic expectations, the less noble metal co-deposits more easily than the more noble one. Anomalous co-deposition occurs frequently in systems such as Fe-Ni, Zn-Fe, and Co-Ni.

Example reaction (Fe-Ni system):

$$Fe^{2+} + 2e^- \rightarrow Fe\ (s) \tag{38}$$

$$Ni^{2+} + 2e^- \rightarrow Ni\ (s) \tag{39}$$

These reactions occur simultaneously, with the alloy composition determined by the relative rates of reduction, influenced by electrolyte conditions and applied potential [72, 73].

## **6.3 Electrodeposition of semiconductors**

Semiconductor electrodeposition is an excellent, low-temperature, cost-effective method for the deposition of thin films and nanostructures employed in optoelectronic, photovoltaic, and sensing devices. In contrast to metallic electrodeposition, featuring simple electron transfer for the reduction of metal cations, semiconductor deposition is inherently more challenging because of the requirement for highly controlled stoichiometry, phase, and crystallinity in order to obtain targeted electronic and optical properties. This process involves co-deposition of cations and anions (frequently from multicomponent electrolytes) or sequential processes to synthesize compound semiconductors such as ZnO, $Cu_2O$, CdTe, PbSe, and SnS.

### *6.3.1 Deposition mechanism*

Semiconductor electrodeposition generally involves the creation of a compound by the controlled reduction at the cathode of cationic and anionic precursors. There are two main approaches employed:

*Single-Bath Co-deposition:* Metal and non-metal precursors are both dissolved in the same electrolyte. A tightly controlled potential is used to reduce both species simultaneously, resulting in the growth of a compound semiconductor on the substrate surface. Precision tuning is needed to match the reduction kinetics of the precursors and prevent unwanted side reactions.

*Two-Step Deposition:* A metal film is initially deposited which is then treated chemically or thermally to react the metal with a non-metal precursor or annealed to create the semiconductor compound. This method is convenient when direct co-deposition is not possible due to contrasting reduction potentials or electrolyte incompatibility.

Example reaction (CdTe deposition):

$$Cd^{2+} + 2e^{-} \rightarrow Cd\ (s) \tag{40}$$

$$TeO_3^{2-} + 4H^{+} + 4e^{-} \rightarrow Te\ (s) + 3H_2O \tag{41}$$

$$Cd + Te \rightarrow CdTe \tag{42}$$

These processes take place at the cathode under potentiostatic control, usually with a reference electrode (such as Ag/AgCl or saturated calomel electrode, SCE) to apply accurate potential and avoid side reaction such as hydrogen evolution or metallic phase formation. Primary process parameters along with pH, temperature, and electrolyte stirring need to be tightly regulated to guarantee stoichiometric deposition, prevent extraneous phases (e.g., anion-rich or metal-rich films), and provide high crystallinity [74–77].

### 6.4 Electrodeposition of nanostructures

Electrodeposition is a very versatile and economic method of fabricating nanostructured materials, such as nanowires, nanotubes, nanoporous films, core-shell structures, and quantum dots, which possess outstanding properties owing to their high surface area-to-volume ratio, adjustable dimensions, and distinctive physicochemical properties. These nanostructures are essential in the fields of energy storage, sensing, biomedicine, nanoelectronics, photocatalysis, and quantum technologies. By using accurate electrochemical control, template-assisted processes, and innovative deposition methods, electrodeposition allows for the production of nanostructures with designed morphologies, compositions, and functionalities. Its low-temperature processing, material versatility, and scalability render it a desirable substitute for high-vacuum processes such as physical vapor deposition or molecular beam epitaxy, especially for mass production of sophisticated materials [78].

## 7. Dimensional control and 2D/3D conformalities in electrodeposition

Dimensional control and conformal coatings in electrodeposition are crucial for the production of high-performance electrochemical devices in fields like energy storage, microelectromechanical systems (MEMS/NEMS), microelectronics, catalysis, and nanotechnology. Dimensional control involves accurate control of film thickness, crystal grain size, and surface morphology at micro- and nanoscales to fulfill certain functional requirements [79]. Equivalently, 2D and 3D conformality guarantees even film deposition over flat surfaces or intricate shapes, for example, thin trenches, vias, pores, or curved substrates, required for sophisticated devices such as semiconductor circuits, battery electrodes, and nanostructured catalysts. All these objectives are fulfilled by a blend of electrochemical kinetics, mass transport optimization, and sophisticated deposition methods, allowing for controlled material properties and scalable fabrication [80, 81].

2D and 3D conformality must be attained to provide even coating of intricate geometries, e.g., high-aspect-ratio trenches or porous substrates. Mass transport constraints in skinny features can result in ion depletion and subsequently thicker deposits at feature edges. Pulse-reverse plating switches cathodic and anodic pulses to redistribute ions and improve penetration into features such as 100 nm-wide vias in microelectronics. Surfactants (e.g., polyethylene glycol) or wetting agents lower the surface tension, enhancing electrolyte contact with hydrophobic or restricted surfaces. Agitation, through stirring or ultrasonication, provides the even ion distribution necessary for coating 3D battery electrodes. Template-assisted electrodeposition with anodized aluminum oxide (AAO) containing 20–200 nm pores or polycarbonate membranes provides conformal synthesis of nanowires and nanotubes, as used in Ni or ZnO nanostructures.

Example reaction (Ni nanowire in AAO template):

$$Ni^{2+} + 2e^{-} \rightarrow Ni\ (s) \tag{43}$$

Conformal Ni nanowires form inside AAO pores in this reaction with diameters equal to the template (e.g., 20–200 nm) [82].

## 8. Parametric variations in electrochemical cells

The operation, efficiency and lifespan of electrochemical cells like batteries, fuel cells, supercapacitors, sensors, and electrolysis cells are controlled by an intricate interplay of parameters that dictate reaction kinetics, charge transport, energy efficiency, and material stability. These parameters need to be optimized to design cell behavior for specific applications ranging from high-power energy storage to high-precision sensing and industrial electroplating [83].

### 8.1 Temperature

Temperature has a significant impact on electrochemical reactions through the modification of ion mobility, reaction kinetics, and properties of the electrolyte. From the Arrhenius equation, increasing temperatures accelerate reactions by lowering activation energy barriers, increasing the kinetics of electrodes, and decreasing solution viscosity for better ion diffusion:

$$k = A.e^{\left(-\frac{E_a}{RT}\right)} \tag{44}$$

where k is the rate constant, A is the pre-exponential factor, $E_a$ is the activation energy, R is the gas constant (8.314 J/mol·K), and T is the absolute temperature in Kelvin [84].

### 8.2 pH and electrolyte composition

The pH of the electrolyte determines ion speciation, electrode potentials, and reaction pathways and thereby has a very strong influence on cell performance. Under acidic conditions (e.g., pH < 3), hydrogen evolution ($2H^+ + 2e^- \rightarrow H_2$) will prevail in electrolysis, and alkaline conditions (pH > 10) promote hydroxide ion reactions, such as oxygen reduction ($O_2 + 2H_2O + 4e^- \rightarrow 4OH^-$) in fuel cells. Buffer solutions, like phosphate or borate buffers, maintain stability of pH to avoid shifts in operation that would change reaction selectivity or induce electrode fouling [85].

### 8.3 Electrode surface area and material

The active surface area of the electrode directly determines the number of available sites for electron transfer, thereby impacting current density and efficiency of reaction. Nanostructured electrodes like porous carbon, metal foams, or nanowire arrays provide much larger surface areas than planar electrodes, which improves performance in devices such as supercapacitors (e.g., activated carbon with >1000 $m^2/g$) or catalysis (e.g., Pt nanoparticles). For instance, porous Ni foam electrode in a battery can enhance current density by 10–100 fold over a flat Ni plate [86].

### 8.4 Applied voltage or current density

The applied current density or voltage determines deposition rates, film morphology, and reaction selectivity. Increased voltages speed up reactions but can initiate side reactions, for example, hydrogen evolution ($2H^+ + 2e^- \rightarrow H_2$), which lowers efficiency. Low current densities (1–10 $mA/cm^2$) in electroplating result in smooth, fine-grained coatings, and high current densities (>50 $mA/cm^2$) can lead to dendritic growth. Pulse electrodeposition with on-off pulses (e.g., 10–100 ms) enhances microstructure and uniformity, as observed in Cu interconnects. Pulse-reverse plating, the use of alternating cathodic and anodic pulses, minimizes stress and improves conformality in high-aspect-ratio features [87].

### 8.5 Cell geometry and configuration

Cell geometry such as electrode spacing, electrolyte volume, and cell design, influences mass transport, internal resistance, and reaction uniformity. Small electrode gaps (e.g., <1 mm) diminish ohmic losses but increase risk of ion depletion, and larger gaps raise resistance. Non-uniform ion distribution in deep or intricate cells causes non-uniform deposition, such as in battery electrodes [88].

### 8.6 Mass transport (diffusion, convection, migration)

Mass transport controls the rate of transport of reactants to the electrode surface and thus affects reaction efficiency. It is achieved through the following:

- *Diffusion:* Concentration gradient-driven diffusion is essential in stationary cells but may be slow and restrict current density.

- *Migration:* Movement of ions in the direction of the electric field increases transport in the case of high-conductivity electrolytes.

- *Convection:* Stirring, ultra-sonication, or forced flow (as in flow batteries) keeps depletion zones low, providing a uniform supply of ions.

Porous electrodes or microfluidic channels facilitate increased mass transport, such as in fuel cells where gas diffusion layers enhance $O_2$ delivery [89].

### 8.7 Pressure (in gaseous electrochemical cells)

Gas pressure affects reaction kinetics and mass transfer in devices such as fuel cells or gas sensors. Increased pressures (e.g., 2–10 atm) enhance gas solubility in the case of gases such as $H_2$ or $O_2$, which boosts reaction rates according to Henry's law. For instance, pressure increase in $H_2$ from 1 to 3 atm in a proton-exchange membrane fuel cell can increase current density by 20–30% [90].

## 9. Electrochemical cells governing nucleation and growth mechanisms

Nucleation and growth mechanisms control the performance and characteristics of materials grown in electrochemical cells. They control how atoms or molecules accumulate on an electrode surface to create stable nuclei (nucleation) and then grow into continuous films, nanostructures, or dendritic structures (growth). The electrochemical cell supplies the driving potential for reduction or oxidation reactions, and its construction and operating conditions regulate the morphology, grain size, porosity, adhesion, and functional properties of deposited material. Nucleation and growth should be understood and optimized to tailor material characteristics for future technologies, such as microelectronics, energy storage, catalysis, and sensors [91].

### 9.1 Nucleation in electrochemical cells

Nucleation is the first step in material formation, when atoms or molecules come together to form stable nuclei that act as seeds for growth. The process of nucleation is responsible for the density, distribution, and size of the nuclei, which have a profound effect on the final microstructure and properties of the material. Nucleation can proceed in two main modes [92]:

(i) Instantaneous nucleation and (ii) progressive nucleation.

The mechanism of nucleation is controlled by various factors:

- *Overpotential:* Increased overpotentials decrease the energy required for nucleation, leading to higher nuclei density and instant nucleation.
- *Substrate surface condition:* Nucleation sites are influenced by surface roughness, cleanliness, and chemical makeup.
- *Electrolyte composition and temperature:* Additives to electrolytes (e.g., surfactants, complexing agents such as citrate) influence the kinetics of nucleation by changing ion adsorption or surface energy.
- *Electrolyte concentration:* Increased concentrations of ions (e.g., 0.5 M $CuSO_4$) enhance the number of depositing species available to lead to faster nucleation [93].

The Scharifker–Hills model (1983) is a technique used to study nucleation mechanisms using current-time transients in chronoamperometry. The model explains the current response during growth and nucleation:

$$I(t)^2 \propto \frac{t}{\left(1+t/t_{max}\right)^2} \tag{45}$$

where I(t) is the current at time t and $t_{max}$ is the time at which the maximum current occurs. The shape of the current-time curve distinguishes instantaneous nucleation (sharp peak and rapid decay) from progressive nucleation (broader peak and slower decay), enabling quantitative analysis of nucleation behavior.

Example reaction (Cu nucleation):

$$Cu^{2+} + 2e^- \rightarrow Cu\ (s) \tag{46}$$

Such a reaction triggers the nucleation process on the cathode, where the substrate ionizability and overpotential dictate whether instantaneous or continuous nucleation prevails [94].

### 9.2 Growth mechanisms

After nucleation, the growth process dictates how nuclei turn into bigger-sized structures, including continuous films, islands, or dendrites. The mode of growth has a key influence on the morphology, mechanical properties, and functionality of the material. Some of the prevalent growth modes are as follows:

- Two-dimensional (2D) layer-by-layer growth (Frank-van der Merwe): The atoms accumulate evenly over the surface, producing smooth, uninterrupted layers. This mode is preferred when the material deposited has good adhesion to the substrate and low surface free energy, such as is the case for Ni plating on well-polished Cu substrates at low current densities [95].

- Three-dimensional (3D) island growth (Volmer-Weber): Nuclei develop into distinct islands that ultimately merge to form a film. This happens when weak substrate interactions between deposited material result in rough or porous films, such as Ag deposition on glass with a seed layer.

- Mixed growth (Stranski-Krastanov): Early 2D layer formation gives way to 3D island formation caused by lattice mismatch or strain, which is typical in semiconductor deposition (e.g., CdTe on ITO) [91].

Example reaction (Ni growth):

$$Ni^{2+} + 2e^- \rightarrow Ni\ (s) \tag{47}$$

## 10. Conclusion

This chapter presents a thorough and multidisciplinary treatment of electrochemical cells, blending core principles with state-of-the-art developments. It starts with half-cell reaction basics, cell potential, and classification of electrochemical systems and proceeds to various applications ranging from traditional batteries and sensors to emerging areas such as $CO_2$ capture, additive manufacturing, and synthesis of biofunctional materials. The treatment of electrochemical reaction mechanisms emphasizes the central role of controlled redox processes in energy conversion and material production. Extensive discussions on electrodeposition, dimensional control, and nucleation/growth mechanisms serve to focus attention on their ability to produce nanostructured and conformal coatings. Furthermore, the effect of operational parameters such as temperature, pH, electrolyte composition, cell geometry, and pressure on system efficiency and performance is treated comprehensively. Through the convergence of old and new ways of thinking, this chapter highlights the revolutionary potential of electrochemical technology in meeting worldwide challenges such as clean energy, green materials, biomedical engineering, and environmental cleanup. Electrochemical cells are still a flexible, scalable, and sustainable technology that spurs scientific innovation and industrial processes, all in pursuit of a sustainable future.

## Author details

Ghulam Nabi[1], Muhammad Rafique[1*], Adnan Khalil[2], Ghulam Nabi[3], Muneeb Irshad[4], Muhammad Adnan Anwar[1], Muhammad Asif Nawaz[1] and Sarmad Masood Shaheen[1]

1 Department of Physics, University of Sahiwal, Sahiwal, Pakistan

2 Institute of Physics, Khwaja Fareed University of Engineering and Information Technology, Rahim Yar Khan, Pakistan

3 Department of Physics, University of Gujrat, Hafiz Hayat Campus, Gujrat, Pakistan

4 Department of Physics, University of Engineering and Technology, Lahore, Pakistan

*Address all correspondence to: mrafique.uet@gmail.com

## References

[1] Hart L. Electrochemistry and Electrochemical Engineering. New York: Library Press; 2018

[2] McNulty RC et al. Understanding the limits of Li-NMC811 half-cells. Journal of Materials Chemistry A. 2023;**11**(34): 18302-18312

[3] Pinaud BA et al. Key considerations for high current fuel cell catalyst testing in an electrochemical half-cell. Journal of the Electrochemical Society. 2017; **164**(4):F321

[4] Kakiuchi T. Salt bridge in electroanalytical chemistry: Past, present, and future. Journal of Solid State Electrochemistry. 2011;**15**: 1661-1671

[5] Özkaya AR. Conceptual difficulties experienced by prospective teachers in electrochemistry: Half-cell potential, cell potential, and chemical and electrochemical equilibrium in galvanic cells. Journal of Chemical Education. 2002;**79**(6):735

[6] Newman J, Tiedemann W. Potential and current distribution in electrochemical cells: Interpretation of the half-cell voltage measurements as a function of reference-electrode location. Journal of the Electrochemical Society. 1993;**140**(7):1961

[7] Burgot J-L, Burgot J-L. Redox reactions and electrochemical cells. Ionic Equilibria in Analytical Chemistry. 2012: 205-227

[8] Zoski CG. Handbook of Electrochemistry. New Mexico, USA: Elsevier; 2006

[9] Xia X et al. Galvanic cell formation in silicon/metal contacts: The effect on silicon surface morphology. Chemistry of Materials. 2000;**12**(6): 1671-1678

[10] Colon M. Novel Encapsulation of Oxidizer Applied to Galvanic Cells: Aluminum/$H_2O_2$ Galvanic Cell as a Case Study. USF Tampa Graduate Theses and Dissertations. 2012. Available from: https://digitalcommons.usf.edu/etd/4017

[11] Rafique M et al. Material and method selection for efficient solid oxide fuel cell anode: Recent advancements and reviews. International Journal of Energy Research. 2019;**43**(7):2423-2446

[12] Parmar M, Balamurugan C, Lee D-W. PANI and graphene/PANI nanocomposite films—Comparative toluene gas sensing behavior. Sensors. 2013;**13**(12):16611-16624

[13] Yuliarto B et al. Enhanced benzene selectivity of mesoporous silica SPV sensors by incorporating phenylene groups in the silica framework. Sensors and Actuators B: Chemical. 2009;**138**(2): 417-421

[14] Gounder Thangamani J et al. Recent advances in electrochemical biosensor and gas sensors based on graphene and carbon nanotubes (CNT)-A. Advanced Materials Letters. 2017;**8**: 196-205

[15] Thangamani G et al. Recent advances in electrochemical biosensor and gas sensors based on graphene and carbon nanotubes (CNT)–A review. Advanced Materials Letters. 2017;**8**(3):196-205

[16] Booth TJ et al. Macroscopic graphene membranes and their extraordinary stiffness. Nano Letters. 2008;**8**(8): 2442-2446

[17] Kuila T et al. Recent advances in graphene-based biosensors. Biosensors and Bioelectronics. 2011;**26**(12): 4637-4648

[18] Lang B. A LEED study of the deposition of carbon on platinum crystal surfaces. Surface Science. 1975;**53**(1): 317-329

[19] Anwar S et al. Synthesis and electrochemical performance of 2D molybdenum carbide (MXene) for supercapacitor applications. Electrochimica Acta. 2025;**512**:145508

[20] Fine GF et al. Metal oxide semi-conductor gas sensors in environmental monitoring. Sensors. 2010;**10**(6): 5469-5502

[21] Pike J et al. Formation of stable $Cu_2O$ from reduction of CuO nanoparticles. Applied Catalysis A: General. 2006; **303**(2):273-277

[22] Srivastava M et al. Additive Manufacturing: Fundamentals and Advancements. Florida: CRC Press; 2019

[23] Biswal T, BadJena SK, Pradhan D. Sustainable biomaterials and their applications: A short review. Materials Today: Proceedings. 2020;**30**:274-282

[24] Jenkins S. Chlorine production from NaCl (chlor-alkali). Chemical Engineering. 2016;**123**(8):39-39

[25] Subramaniam A, Schwartz DT, Subramanian VR. Electrochemical modeling and simulation of battery systems: Approaches for design, control, and multiscale simulations. In: Electrochemical Society Meeting Abstracts 239. United states of America: The Electrochemical Society, Inc; 2021

[26] Lakshmanan S, Murugesan T. The chlor-alkali process: Work in progress. Clean Technologies and Environmental Policy. 2014;**16**:225-234

[27] Shim J-H, Jeong J-Y, Park J-Y. Effects of operating conditions on improving alkali and electrical efficiency in chlor-alkali diaphragm cell. International Journal of Electrochemical Science. 2015; **10**(8):6338-6347

[28] Ryu J-S et al. Performance evaluation of diaphragm electrolysis cell for alkali production. Desalination and Water Treatment. 2016;**57**(51): 24697-24703

[29] Hou M et al. A clean and membrane-free chlor-alkali process with decoupled $Cl_2$ and $H_2$/NaOH production. Nature Communications. 2018;**9**(1):438

[30] Wang Y et al. Ti-substituted tunnel-type $Na_{0.44}MnO_2$ oxide as a negative electrode for aqueous sodium-ion batteries. Nature Communications. 2015; **6**(1):6401

[31] Cao Y et al. Reversible sodium ion insertion in single crystalline manganese oxide nanowires with long cycle life. Advanced Materials. 2011;**23**(28): 3155-3160

[32] Sauvage F et al. Study of the insertion/deinsertion mechanism of sodium into $Na_{0.44}MnO_2$. Inorganic Chemistry. 2007;**46**(8):3289-3294

[33] Whitacre J, Tevar A, Sharma S. $Na_4Mn_9O_{18}$ as a positive electrode material for an aqueous electrolyte sodium-ion energy storage device. Electrochemistry Communications. 2010;**12**(3):463-466

[34] Saint JA, Doeff MM, Wilcox J. Electrode materials with the $Na_{0.44}MnO_2$ structure: Effect of titanium substitution on physical and electrochemical properties.

Chemistry of Materials. 2008;**20**(10): 3404-3411

[35] Renfrew SE, Starr DE, Strasser P. Electrochemical approaches toward $CO_2$ capture and concentration. ACS Catalysis. 2020;**10**(21):13058-13074

[36] Liu T et al. Continuous decoupled redox electrochemical $CO_2$ capture. Nature Communications. 2024;**15**(1): 10920

[37] Li N et al. Understanding of electrochemical mechanisms for $CO_2$ capture and conversion into hydrocarbon fuels in transition-metal carbides (MXenes). ACS Nano. 2017; **11**(11):10825-10833

[38] Bruggeman D, Rothenberg G, Garcia A. Investigating proton shuttling and electrochemical mechanisms of amines in integrated $CO_2$ capture and utilization. Nature Communications. 2024;**15**(1):9207

[39] An X et al. Application of ionic liquids in $CO_2$ capture and electrochemical reduction: A review. Carbon Resources Conversion. 2023; **6**(2):85-97

[40] Gibson I et al. Additive Manufacturing Technologies. Vol. 17. Switzerland: Springer; 2021

[41] Basha MM et al. State of the art on chemical and electrochemical based finishing processes for additive manufactured features. Additive Manufacturing. 2022;**58**:103028

[42] Ajmal S, Hashmi FA, Imran I. Recent progress in development and applications of biomaterials. Materials Today: Proceedings. 2022;**62**:385-391

[43] Grimshaw J. Electrochemical Reactions and Mechanisms in Organic Chemistry. Netherlands: Elsevier; 2000

[44] Fomo G et al. Electrochemical polymerization. Functional Polymers. 2019:105-131

[45] Cho SI, Lee SB. Fast electrochemistry of conductive polymer nanotubes: Synthesis, mechanism, and application. Accounts of Chemical Research. 2008;**41**(6):699-707

[46] Tajik S et al. Recent developments in conducting polymers: Applications for electrochemistry. RSC Advances. 2020; **10**(62):37834-37856

[47] Besenhard JO. Handbook of Battery Materials. Germany: John Wiley & Sons; 2008

[48] Barak M. Electrochemical Power Sources: Primary and Secondary Batteries. England: IET; 1980

[49] Goodenough JB. Electrochemical energy storage in a sustainable modern society. Energy & Environmental Science. 2014;**7**(1):14-18

[50] Liu L, Zhu J, Zheng L. An effective method for estimating state of charge of lithium-ion batteries based on an electrochemical model and Nernst equation. IEEE Access. 2020;**8**: 211738-211749

[51] Jiang J et al. The influence of electrolyte concentration and solvent on operational voltage of Li/CFx primary batteries elucidated by Nernst equation. Journal of Power Sources. 2022;**527**: 231193

[52] Larcher D, Tarascon J-M. Towards greener and more sustainable batteries for electrical energy storage. Nature Chemistry. 2015;**7**(1):19-29

[53] Li M et al. 30 years of lithium-ion batteries. Advanced Materials. 2018; **30**(33):1800561

[54] Kondo Y, Abe T, Yamada Y. Kinetics of interfacial ion transfer in lithium-ion batteries: Mechanism understanding and improvement strategies. ACS Applied Materials & Interfaces. 2022;**14**(20): 22706-22718

[55] Wu X et al. In situ transmission electron microscopy studies of electrochemical reaction mechanisms in rechargeable batteries. Electrochemical Energy Reviews. 2019;**2**:467-491

[56] Muench S et al. Polymer-based organic batteries. Chemical Reviews. 2016;**116**(16):9438-9484

[57] Zubi G et al. The lithium-ion battery: State of the art and future perspectives. Renewable and Sustainable Energy Reviews. 2018;**89**:292-308

[58] Garche J et al. Encyclopedia of Electrochemical Power Sources. Netherlands: Newnes; 2013

[59] Bagotsky VS, Skundin AM, Volfkovich YM. Electrochemical Power Sources: Batteries, Fuel Cells, and Supercapacitors. Canada: John Wiley & Sons; 2015

[60] Cooper A, Lam LT, Moseley PT, Rand DA. The next great challenge for valve-regulated lead-acid batteries: High-rate partial-state-of-charge duty in new-generation road vehicles. In: Valve-Regulated Lead-Acid Batteries. Elsevier. 2004. pp. 549-565

[61] Huggins RA, Huggins RA. Energy Storage: Fundamentals, Materials and Applications. Switzerland: Springer; 2016

[62] ul Haque S et al. A review: Evolution of enzymatic biofuel cells. Journal of Environmental Management. 2021;**298**: 113483

[63] Zhou J et al. Research progresses and application of biofuel cells based on immobilized enzymes. Applied Sciences. 2023;**13**(10):5917

[64] Bojang AA, Wu HS. Characterization of electrode performance in enzymatic biofuel cells using cyclic voltammetry and electrochemical impedance spectroscopy. Catalysts. 2020; **10**(7):782

[65] Calabrese Barton S, Gallaway J, Atanassov P. Enzymatic biofuel cells for implantable and microscale devices. Chemical Reviews. 2004;**104**(10): 4867-4886

[66] Wang L et al. Enzymatic biofuel cell: Opportunities and intrinsic challenges in futuristic applications. Advanced Energy and Sustainability Research. 2021;**2**(8): 2100031

[67] Chae W et al. A brief review of gel polymer electrolytes using in situ polymerization for lithium-ion polymer batteries. Polymers. 2023;**15**(4):803

[68] Zhao X et al. A review of polymer-based solid-state electrolytes for lithium-metal batteries: Structure, kinetic, interface stability, and application. Batteries & Supercaps. 2023;**6**(4): e202200502

[69] Mecerreyes D et al. Current trends and perspectives of polymers in batteries. Macromolecules. 2024;**57**(7): 3013-3025

[70] Liu F et al. Electrodeposition of metals and alloys from ionic liquids. Journal of Alloys and Compounds. 2016; **654**:163-170

[71] Zhong Z et al. A review of external field-enhanced metal electrodeposition: Mechanism and applications. The

Journal of the Minerals. 2025;**77**(2):665-685

[72] Al-Duaij O, Abou-Krisha M, Attia M. Influence of the deposition temperature on the electrodeposition mechanism of Zn-Co-Fe alloy. International Journal of Electrochemical Science. 2017;**12**(12):11972-11986

[73] Omar IM, Emran KM, Aziz M. Electrodeposition of Ni-Co film: A review. International Journal of Electrochemical Science. 2021;**16**(1):150962

[74] Pandey RK, Sahu S, Chandra S. Handbook of Semiconductor Electrodeposition. Florida: CRC Press; 2017

[75] Kowalik R, Kazimierczak H, Żabiński P. Electrodeposition of cadmium selenide. Materials Science in Semiconductor Processing. 2016;**50**:43-48

[76] Ojo AA, Dharmadasa I. Factors affecting electroplated semiconductor material properties: The case study of deposition temperature on cadmium telluride. Coatings. 2019;**9**(6):370

[77] Chen J, Wan C. Dependence of the composition of CdTe semiconductor on conditions of electrodeposition. Journal of Electroanalytical Chemistry. 1994;**365**(1–2):87-95

[78] Nasirpouri F. Electrodeposition of Nanostructured Materials. Vol. 62. Switzerland: Springer; 2017

[79] Rafique M, Tahir MB, Anwar S. MXenes for Energy Storage Applications: Emerging Characteristics, Compositions, and Synthesis Methods. Florida: CRC Press; 2024

[80] Maciulis V, Ramanaviciene A, Plikusiene I. Recent advances in synthesis and application of metal oxide nanostructures in chemical sensors and biosensors. Nanomaterials. 2022;**12**(24):4413

[81] Hou L. Advanced 3D Microfabrication and Demonstration of Arrayed Electrowetting Microprisms. United states of America: University of Cincinnati; 2012

[82] Braun TM, Schwartz DT. The emerging role of electrodeposition in additive manufacturing. The Electrochemical Society Interface. 2016;**25**(1):69

[83] Baumann M et al. Parameter variations within Li-ion battery packs–theoretical investigations and experimental quantification. Journal of Energy Storage. 2018;**18**:295-307

[84] Alipour M et al. A review on temperature-dependent electrochemical properties, aging, and performance of lithium-ion cells. Batteries. 2020;**6**(3):35

[85] Periyasamy S, Muthuchamy M. Electrochemical oxidation of paracetamol in water by graphite anode: Effect of pH, electrolyte concentration and current density. Journal of Environmental Chemical Engineering. 2018;**6**(6):7358-7367

[86] Vengatesan S, Venkadesh R, Khan MW. Parametric study on geometric and material properties of polymer electrolyte membrane fuel cell with free vibration analysis. International Journal of Hydrogen Energy. 2023;**48**(47):18022-18031

[87] Deconinck J. Current Distributions and Electrode Shape Changes in Electrochemical Systems. Vol. 75. Germany: Springer Science & Business Media; 2012

[88] Arenas L, De León CP, Walsh F. Critical review—The versatile plane

parallel electrode geometry: An illustrated review. Journal of the Electrochemical Society. 2020;**167**(2): 023504

[89] Becker M, Turek T. New mass transport correlation for vanadium redox-flow batteries based on a model-assisted parameter estimation. Batteries. 2023;**9**(5):253

[90] Chen H, Wang J, Xu X. Parametric study of operating conditions on performances of a solid oxide electrolysis cell. Journal of Thermal Science. 2023; **32**(6):1973-1988

[91] Dubrovskii VG. Nucleation Theory and Growth of Nanostructures. Germany: Springer; 2014

[92] Nasirpouri F, Nasirpouri F. An overview of nanostructured materials. Electrodeposition of Nanostructured Materials. 2017;**62**:1-42

[93] Sounart TL et al. Sequential nucleation and growth of complex nanostructured films. Advanced Functional Materials. 2006;**16**(3): 335-344

[94] Lin J et al. Understanding the nanoscale phenomena of nucleation and crystal growth in electrodeposition. Nanoscale. 2024;**16**(42):19564-19588

[95] Rafique M et al. Novel synthesis and electrochemical performance of 2D molybdenum nitride (MXene) for supercapacitor application. Journal of Energy Storage. 2024;**103**:114354

Chapter 2

# Self-Assembly in Thin-Film Photovoltaics: Foundations and Mechanisms

*Zhenyu Zhang, Zhu Ma, Zhuowei Du, Yi Chen, Qian Zhang, Hao Du, Zhuo Lv, Siming Huang, Maryam Shahrostami, Javad Maleki, Narges Yaghoobi Nia, Mojtaba Abdi-Jalebi and Mahmoud Zendehdel*

## Abstract

Thin-film photovoltaics have emerged as a promising technology for sustainable energy conversion, driven by their low-cost fabrication, lightweight design, and versatility. This chapter explores the pivotal role of self-assembly processes in the development of high-performance thin film photovoltaic devices. Self-assembly, a bottom-up fabrication approach, enables precise control over material organization at the nanoscale, resulting in optimized light absorption, charge transport, and overall device efficiency. In this chapter, we outline the fundamental principles governing molecular self-assembly, including noncovalent interactions, packing motifs, and thermodynamic driving forces, and elucidate their relevance to solar energy conversion. We further examine the design criteria for self-assembled monolayers (SAMs), coordination polymers, and supramolecular architectures in modulating surface energetics, charge selectivity, and interfacial stability in thin-film PVs. By drawing parallels to biological self-organization, the chapter highlights how synthetic systems can emulate nature's efficiency in capturing and transferring energy. This foundational understanding sets the stage for rationally designing self-assembling molecules and interfaces, thereby accelerating the emergence of scalable, stable, and high-efficiency solar energy technologies.

**Keywords:** self-assembly monolayer, thin-film photovoltaics, surface energy modulation, noncovalent interactions, molecular packing, nanoscale engineering, light management, charge transport, charge extraction, solvent engineering, supramolecular architectures, sustainable energy conversion, chemical bonding, molecular orientation, chemical bath deposition, zero-waste coating process

# 1. Introduction

## 1.1 Development status of thin-film photovoltaics

With the growing global demand for clean energy, solar power has gained significant attention. Silicon-based solar cells dominate the market due to their compatibility with mature microelectronics technology, balancing cost and performance. However, silicon's high material requirements and poor light absorption necessitate thick substrates. To address these challenges, thin-film photovoltaic technology has emerged as a promising alternative, offering advantages like lower cost, lightweight design, and flexibility, making it a key focus in solar energy research [1]. In recent years, the power conversion efficiency of thin-film photovoltaic devices has been continuously improved, but there is still a certain gap from the theoretical limit. Therefore, it is necessary to continuously explore new technologies and methods to further enhance their performance.

## 1.2 Significance of introducing self-assembly technology

Self-assembly technology presents a novel approach to advancing thin-film photovoltaics. Over the past three decades, nanomaterials have gained increasing attention from both academia and industry. Their precise nanoscale manipulation through self-assembly offers significant potential for improving the performance of thin-film photovoltaic devices [2]. At the nanoscale, materials exhibit significantly altered physical and chemical properties. Self-assembly technology enables precise control over the microstructure of thin-film photovoltaic materials through techniques like intermolecular interactions, templating, and solvent engineering. For example, in the light-absorbing layer, self-assembly can create nanoporous structures or quantum dot superlattices, enhancing device performance [3]. The nanoporous structure can increase the scattering and reflection of light within the material, prolong the light propagation path, and thus improve the light absorption efficiency. The quantum dot superlattice structure, on the other hand, can achieve efficient absorption of light at different wavelengths through the coupling effect between quantum dots, broadening the light absorption range [4]. In terms of charge transport, the ordered nanochannels or layered structures formed by self-assembly provide rapid paths for charge transfer, reduce charge recombination, and improve the charge collection efficiency, thereby enhancing the power conversion efficiency of photovoltaic devices.

From the perspective of the preparation process, self-assembly technology has unique advantages. Traditional thin-film preparation methods often suffer from complex processes and high costs [5]. Self-assembly is spontaneous and adaptable, requiring minimal external intervention. During solution spin-coating, material molecules naturally form specific structures as the solvent evaporates, simplifying fabrication, reducing costs, and minimizing process steps. Additionally, *in situ* growth through self-assembly prevents impurities and defects, enhancing material quality and stability [6].

Self-assembly technology opens new avenues for innovation in thin-film photovoltaics, enabling the development of novel materials and structures. For instance, it allows the integration of organic and inorganic materials to create hybrid photovoltaic materials, combining the flexibility of organics with the high carrier mobility of inorganics, resulting in superior optoelectronic properties [7]. Self-assembly technology also enables the multifunctionalization of thin-film photovoltaic devices. By incorporating self-assembled nanostructures with sensing capabilities, intelligent devices can be developed that combine power conversion with sensing functions. Self-assembly technology plays a crucial role in thin-film photovoltaics, enhancing performance, streamlining fabrication, and driving innovation. As research advances, it will further boost the industry, accelerate solar energy adoption, and contribute to global sustainable energy development.

### 1.3 Basic principles of self-assembly

#### *1.3.1 Molecular interactions*

Supramolecular chemistry, dealing with noncovalent interactions, is one of the frontiers of twenty-first-century chemistry. Originally developed by Jean-Marie Lehn, Donald Cram, and Charles Pedersen, the field is defined as "the chemistry of molecular assembly and intermolecular interactions". Self-assembly refers to the process in which molecules, nanoparticles, or other structural units spontaneously organize to form well-defined structures without external intervention. This process is often driven by forces and interactions between molecules. The main noncovalent forces involved are van der Waals forces, hydrogen bonds, and electrostatic interactions [8].

##### *1.3.1.1 Van der Waals forces*

Van der Waals forces, including orientation, induction, and dispersion forces, play a key role in the self-assembly of thin-film photovoltaic materials. In organic small-molecule photovoltaics, dispersion forces drive molecules to arrange in an ordered structure. As solution concentration increases or the solvent evaporates, these forces become more pronounced, guiding molecular aggregation and facilitating the formation of a stable, orderly film structure. Zuo et al. [9] applied 3-alanine molecules to self-assembly materials (SAM) (**Figure 1a**). AFM (**Figure 1b-c**) and SEM (**Figure 1d-e**) analyses revealed that the plate-like crystalline domains of the modified SAM material expanded laterally compared to the rod-like crystalline domains in the perovskite solar cell (PSC) without SAM. In addition, SAM also induced the transition of morphology from rod-like to plate-like. All the XRD peaks corresponding to the $CH_3NH_3PbI_3$ perovskite film were enhanced upon the insertion of the SAM, which was indicative of improved crystallinity (**Figure 1f**).

##### *1.3.1.2 Hydrogen bonds*

Hydrogen bonding, a highly directional and selective intermolecular interaction, plays a crucial role in polymeric photovoltaic materials. It influences molecular chain conformation and promotes interchain connections, forming ordered supramolecular structures. For instance, in poly(3-hexylthiophene) (P3HT) and the fullerene derivative PCBM, hydrogen bonding facilitates self-assembly in solution, creating microstructures that enhance charge separation and transport [10, 11].

##### *1.3.1.3 Electrostatic interactions*

Electrostatic interactions, arising from the attraction or repulsion between charged molecules or ions, play a crucial role in regulating self-assembly in thin-film

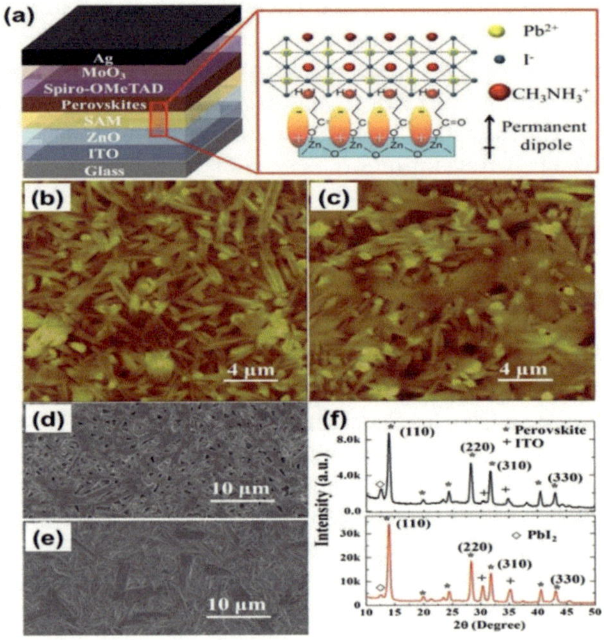


**Figure 1.**
*The representative study stresses the importance of morphology control through surface modification with SAM to achieve the desirable device performance of the PSC. (a) A schematic diagram of a PSC based on the SAM-modified ZnO layer. AFM image of the PSC on (b) bare ZnO surface and (c) the SAM-modified ZnO surface. SEM images of the PSC on (d) bare ZnO surface and (e) the SAM-modified ZnO surface. (f) XRD patterns of perovskite films on ZnO (black line) and ZnO/SAM substrates (red line). Note that the dipole direction in (a) obeys the chemistry definition, which is often opposite to the one in physics. Reproduced with permission. Copyright (2015), American Chemical Society.*

photovoltaics. In quantum dot solar cells, controlling the charge properties and concentration on the quantum dot surface enables electrostatic-driven self-assembly into ordered arrays on substrates or between organic ligands. These ordered arrays enhance charge carrier transport and collection efficiency, improving device performance. Yang et al. [12] covalently attached a polar passivation group, hydroxyl (-OH), to a polar solvent-proof hydrophobic hole transporting polymer, enabling the bottom-surface passivation of perovskite layer for efficient p-i-n type PSCs. The -OH is linked to a long alkyl chain-substituted conjugated polymer backbone via a short methylene unit, endowing the resultant amphiphilic polymer (P15, **Figure 2**) with a strong resistance to the polar solvents. To avoid the shielding effect of hydrophobic segments on polar passivation groups during the formation of the P15 film, a solvent evaporation-induced self-assembly method is used to control the distribution of -OH groups on the film surface. By adjusting solvent polarity, P15 can form “crew-cut” micelle-like aggregates through dip-coating or homogeneous films with -OH enriched on the surface via spin-coating. This allows the polar perovskite precursor solution to spread evenly on the P15 film without dewetting or damaging the underlying layer, resulting in a high-quality perovskite layer. The exposed -OH groups on the P15 surface passivate perovskite layer defects by interacting with undercoordinated Pb2+ and I-, reducing trap density. Consequently, a p-i-n type perovskite solar cell (PSC) based on P15 achieves a remarkable power conversion efficiency (PCE) of 20.12%, outperforming devices based on reference polymers without -OH and the classical PTAA.

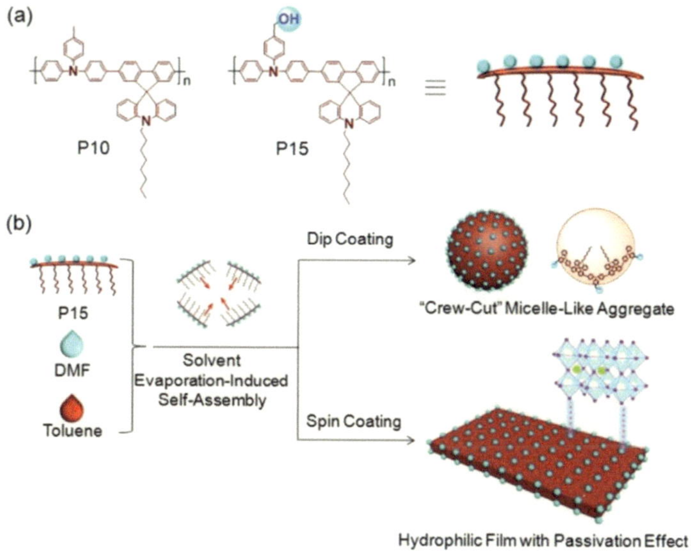


**Figure 2.**
*(a) Chemical structures of P10 and P15. (b) Illustration of self-assembly processes of P15 by dip-coating and spin-coating.12 Copyright (2021), Wiley-VCH GmbH.*

The P15-based device also shows advantages in reproducibility, large-area fabrication, and photostability compared to PTAA.

### *1.3.2 Template technology*

#### *1.3.2.1 Hard templates*

Hard templates are usually rigid materials with specific shapes and sizes, such as porous alumina templates, mesoporous silica templates, etc. [13] In thin-film PV self-assembly, hard templates provide precise spatial constraints for material growth and assembly. Taking the porous alumina template as an example, it has a highly ordered nanoscale pore structure. The precursor solution of photovoltaic materials is introduced into these channels, and nanostructures, such as nanowires and nanotubes, which are consistent with the shape of the channels, can be formed in the channels through chemical deposition and electrochemical deposition [14]. These nanostructures are highly ordered and homogeneous and can be used as efficient light absorption and charge transport channels in thin-film photovoltaic devices. Guo et al. [15] propose a light-induced conformation conversion strategy to construct a rigid polymer skeleton by transforming its aromatic configuration into a rigid quinone configuration, as confirmed by Fourier transform infrared (FT-IR) spectroscopy and Raman spectroscopy (**Figure 3**). X-ray diffraction and morphology characterization revealed that the quinone conformation enhances molecular π–π stacking and aggregation in solid films, improving charge transport. This leads to better device efficiency in PM6:L8-BO and D18:L8-BO OSCs, with improved fill factor (FF) and short-circuit current density ($J_{SC}$). The method demonstrates versatility across photovoltaic systems, achieving a maximum PCE of 19.9% in the D18:PM6:L8-BO ternary device and reaching a milestone PCE of 20%.

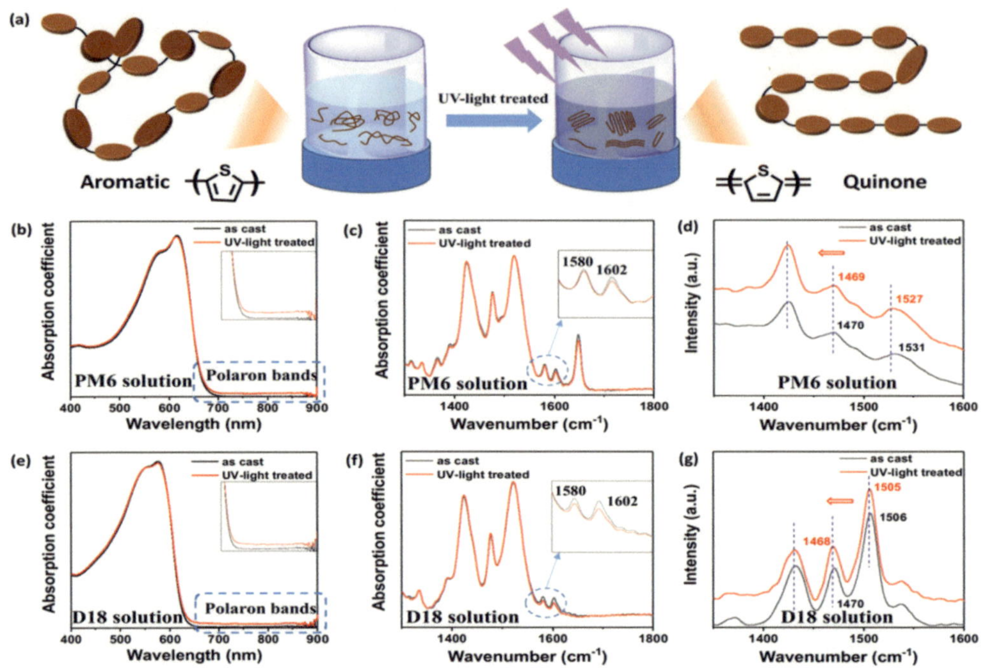


**Figure 3.**
*(a) Schematic diagram of light-induced pre-aggregation of a polymer, where the solution vial was placed invertedly with a UV light cast through the top. (b) Absorption spectra, (c) Fourier transform infrared (FT-IR) spectra, and (d) Raman spectra of the PM6 solution obtained with different treatments. (e) Absorption spectra, (f) FT-IR spectra, and (g) Raman spectra of D18 solution obtained with different treatments [15]. Copyright (2024), Energy Environmental Science.*

#### *1.3.2.2 Soft Templates*

Soft templates, like micelles and liquid crystals, are formed by self-assembly of surfactants and polymers in solution and have flexible, adjustable structures. In thin-film photovoltaics, these templates guide PV material molecules to assemble around or within them. For example, micelles can encase precursor materials, allowing them to react and self-assemble into nanoparticles, with size and shape controlled by adjusting the soft template's composition and structure. Tong et al. [16] designed an effective asymmetric $\pi$-extended SAM, (4-(9H-dibenzo[a,c]carbazol-9-yl)butyl)phosphonic acid (A-4PADCB), which exhibits optimal coverage on ITO and enhanced interface contact with perovskite. This design improved hole extraction, reduced interfacial non-radiative recombination, and lowered defect density. The asymmetric structure, with a dipole orientation away from the center, regulated the tilt angle between the terminal benzene ring and the surface normal, promoting better growth of the upper perovskite layer (**Figure 4**). Implementing A-4PADCB as an SBS for inverted PSCs yielded impressive PCEs of 25.05% for f-PSCs and 20.64% (certified 19.51%) for flexible perovskite modules (f-PSMs) with an aperture area of 20.84 $cm^2$. Additionally, f-PSCs based on A-4PADCB retained 85% of their initial efficiency following prolonged heat testing at 85°C for 500 h and 87% of their original PCE after 400 h of maximum power-point tracking (MPPT) under continuous illumination. The enhanced mechanical reliability of perovskite films resulted in f-PSCs exhibiting superior mechanical stability, retaining 93% of their initial efficiency after 5000 bending cycles at a radius of 5 mm.

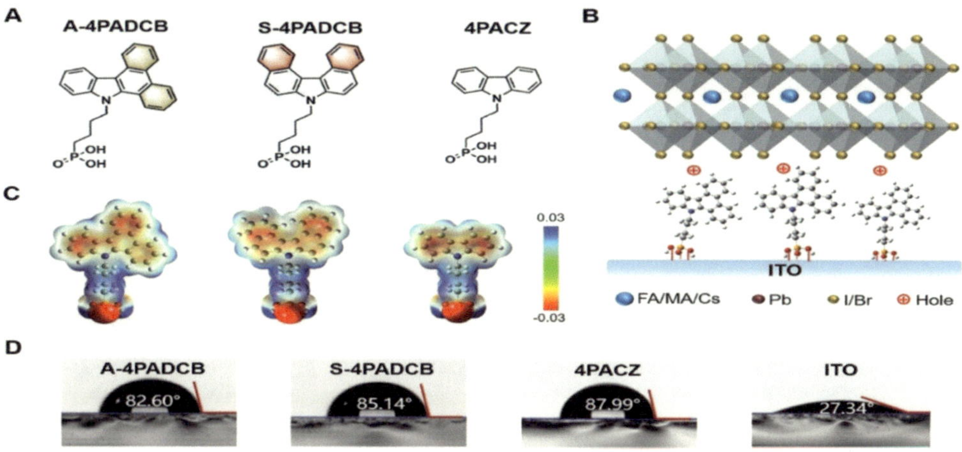


**Figure 4.**
*Material properties of three SAMs and Schematic of SAMs as HSL between ITO and perovskite. (a) Molecular structure of A-4PADCB, S-4PADCB, and 4PACZ. (b) Schematic of A-4PADCB as HSL between ITO and perovskite. (c) Calculated ESP distributions of A-4PADCB, S-4PADCB, and 4PACZ. (d) Water contact angle of SBSs and bared ITO [16]. Copyright (2024), Wiley-VCH GmbH.*

### *1.3.3 Solvent engineering*

#### *1.3.3.1 Influence of solvent selection*

Solvent properties significantly affect the self-assembly of thin-film photovoltaic materials. Polarity, solubility, and boiling point influence molecular interactions and assembly kinetics. For example, in polar solvents, strong solute-solvent interactions promote dispersion, while in non-polar solvents, intermolecular forces drive aggregation and self-assembly. Yang et al. [12] studied the self-assembly behavior of conjugated amphiphilic P15 by adjusting the mixed ratio of toluene and DMF. By considering solubility and solvent selectivity of the -OH, a polar solution (1.5 mg $mL^{-1}$) with a volume ratio of DMF/toluene = 4/1 was adopted. To visualize these self-assembled structures, transmission electron microscopy (TEM) was carried out after drying the same solution on the TEM grid at room temperature. Highly polydispersed solid spheres with diameters between 100 and 800 nm are clearly presented, and some of them adhere or aggregate together (**Figure 5a-c**). A mechanism for solvent evaporation-induced self-assembly of P15 is proposed. In a DMF-rich solution, solvophobic forces drive P15 to form micelles with hydrophobic alkyl chains inward and hydrophilic -OH groups exposed. As toluene evaporates faster than DMF, the environment becomes more polar, favoring -OH interactions while forcing alkyl chains to aggregate into hydrophobic cores with -OH coronas, forming "crew-cut" micelle-like structures. Further evaporation leads to cohesive spheres stabilized by hydrogen bonding and the rigid conjugated skeleton of P15. Spin-coating generates a P15 film with exposed - OH groups. The elevated contact angle indicates more hydroxyl functionalities on the film surface. These hydroxyl groups interact with undercoordinated $Pb^{2+}$ and $I^{-}$ sites, passivating surface defects in the upper perovskite layer. This enables perovskite precursor deposition without damaging the underlying layer (**Figure 5d-f**).

In addition, the boiling point of the solvent will also affect the self-assembly process. The evaporation speed of the solvent with lower boiling point is fast, which may

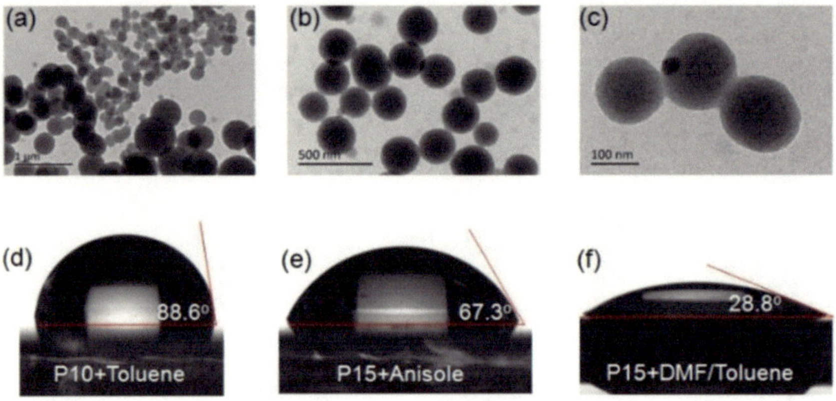


**Figure 5.**
*(a–c) TEM images of P15 by dip-coating. (d–f) water contact angles of polymer films by spin-coating [12]. Copyright (2021), Wiley-VCH GmbH.*

lead to the self-assembly process too fast and it is difficult to form the ideal ordered structure. However, solvents with higher boiling points volatilize slowly, which is beneficial for the molecules to have enough time to arrange in an orderly manner.

### *1.3.3.2 Solvent evaporation-induced self-assembly*

Solvent volatilization is crucial in film preparation, driving self-assembly by increasing solute concentration and enhancing intermolecular interactions. During spin-coating, rapid solvent evaporation accelerates molecular interactions, such as van der Waals forces and hydrogen bonding, leading to structured thin films. Precise control of evaporation rate, temperature, and humidity allows tuning of film morphology. For instance, while DMSO co-solvent and anti-solvent dripping are common in Sn-based perovskite film fabrication, DMSO oxidizes $Sn^{2+}$, and rapid dripping induces defects, compromising film stability. Based on the above results, Dong et al. [17] used HAAc to improve the crystallization process of $SnI_2$/FAI precursor solution, and the model is shown in (**Figure 6**). In the absence of HAAc, the $SnI_2$/FAI/DMF coordination complex forms colloidal clusters that slowly transform into a polyhalide framework. Weak $SnI_2$-FAI interactions lead to slow nucleation during spin-coating, resulting in discontinuous films with isolated crystals. With HAAc, micelle-assisted assembly forms a dense film with oriented particles. Based on the LaMer model, HAAc lowers the nucleation barrier, inducing high-density pre-formed crystals that self-assemble into ordered films (**Figure 6a,b**). This method overcomes the nucleation-growth coupling issue in anti-solvent deposition, enabling DMSO-free, anti-solvent spin-coating of Sn-based perovskite films.

## 1.4 Self-assembly customized nano-ordered structure

### *1.4.1 Formation of nanoscale ordered structures*

Advancing photovoltaic technology requires continuous improvements in film performance. Customized self-assembled nano-ordered structures offer a promising approach to enhancing photoelectric conversion efficiency, stability, and other key properties. These nanoscale structures precisely control light-matter interactions and

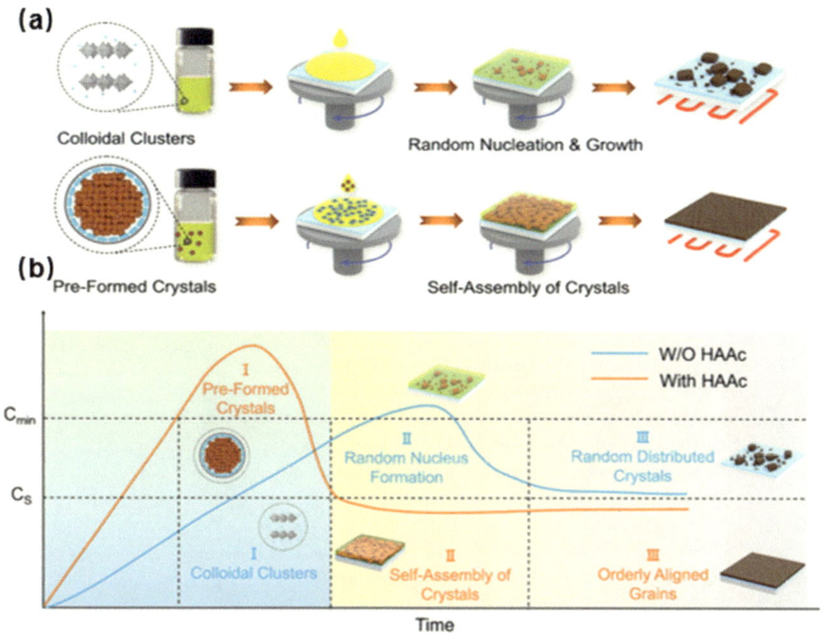


**Figure 6.**
*(a) Schematic of film-forming process based on precursor solution without and with HAAc. (b) Schematic of the LaMer model for demonstrating the different stages of crystallization without and with HAAc [17]. Copyright (2025), Wiley-VCH GmbH.*

charge carrier transport, demonstrating significant potential for photovoltaic applications. **Figure 7** gives a detailed overview of the current research work centered on interfacial self-assembled nanostructures, including their basic principles, construction methods, and their applications in various fields.

#### *1.4.1.1 Precise regulation of nanostructure size and morphology*

Self-assembly technology offers distinct advantages in fabricating nanostructured electrodes with high surface area and conductivity, such as nanowire arrays and nanoporous electrodes. For instance, highly ordered zinc oxide nanowire array electrodes can be grown using template-assisted self-assembly. This involves preparing an ordered porous template, introducing a zinc oxide precursor solution, and forming nanowires through chemical reactions and heat treatment. These nanowire array electrodes enhance light absorption and improve photogenerated charge carrier collection efficiency. Guo et al. [19] investigate the effects of a family of carbazole derivative-based SAMs as the nonpolymeric hole selective contacts on the crystal grown models, spatial multiscale structures as well as the formed energy landscape of MA-free RP-phase 2D perovskite films. They discover that such SAM layer prefers to align in ordered lateral orientation with larger crystal domains, which facilitates the following enhanced crystal growth of 2D perovskites. More importantly, the 2D perovskite films grown on SAM layers present a spatially ordered 2D/3D heterogeneity in long range, where the diverse 2D phase ($n$ = 2, and 3) domains are visualized to be tightly coupled in short range (**Figure 8**). These rationally designed heterostructures create an optimized energy landscape with ordered interface contacts, enhancing radiative exciton recombination while suppressing nonradiative losses. This SAM-based

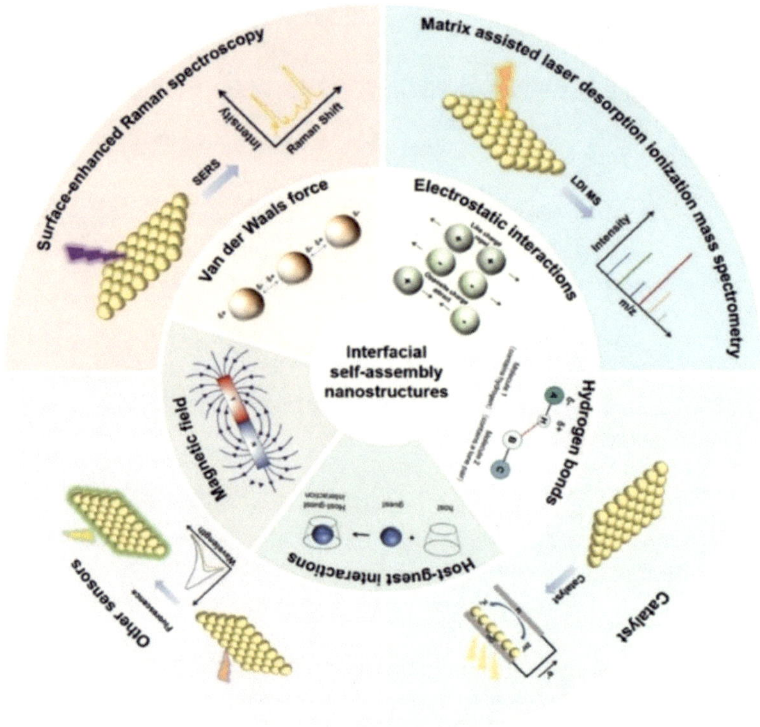


**Figure 7.**
*An overview of the construction methods and applications of interfacial self-assembly nanostructures [18]. Copyright (2024), Wiley-VCH GmbH.*

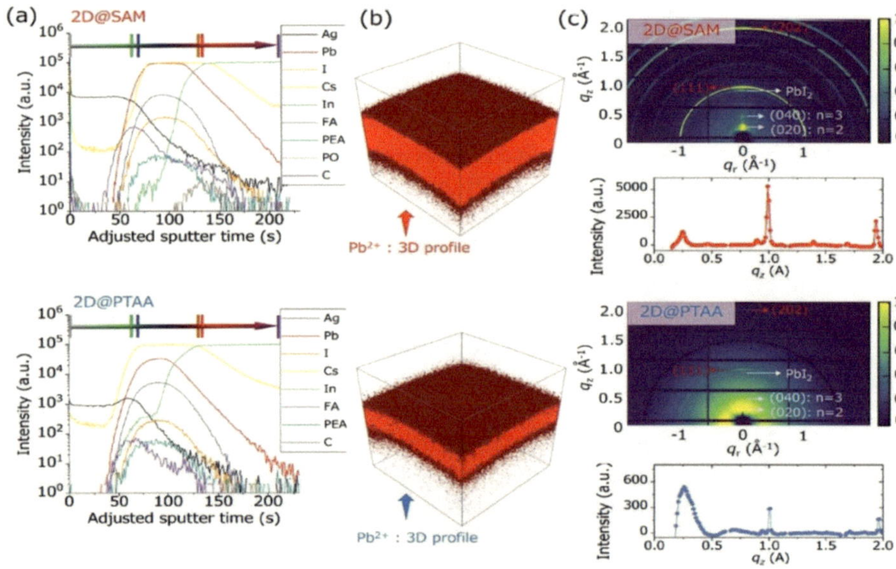


**Figure 8.**
*Long-ranger order of 2D perovskites grown on PTAA and SAM (2PACz). (a) TOF-SIMS depth profiles of 2D SAM and 2D PTAA-based full devices. (b) TOF-SIMS 3D spatial profile mappings of $Pb^{2+}$ signal in 2D SAM and 2D PTAA. (c) GIWAXS patterns and corresponding out-of-plane line cuts of the 2D SAM and 2D PTAA [19]. Copyright (2024), Wiley-VCH GmbH.*

strategy is applicable to various 2D perovskite prototypes, scalable areas, and flexible devices. Notably, they achieve a champion efficiency of 18.85% (certified 18.19%) for RP-phase 2D perovskite (n = 5) solar cells using selective SAMs, demonstrating superior damp-heat and operational stability compared to PTAA-based devices.

#### *1.4.1.2 Achievement of highly ordered molecular arrangement*

Self-assembly is a simple and cost-effective process for fabricating simple components into ordered structures, including nanowires (NWs), nanotubes, nanopins, and nanoarrays [13, 20]. Self-assembly can enhance the conductivity and stability of electrode materials. By integrating conductive nanomaterials like carbon nanotubes or graphene, composite electrodes with significantly improved conductivity can be formed. Additionally, the self-assembly process creates strong bonding between the electrode material and the substrate, enhancing stability. For instance, electrodes fabricated by self-assembling graphene with polymers exhibit high conductivity, stability, and flexibility, effectively boosting the performance of thin-film photovoltaic devices.

### *1.4.2 Enhancement of photoelectric performance*

#### *1.4.2.1 Significant improvement of light absorption performance*

In thin-film photovoltaic devices, self-assembled nanoscale structures, such as nanowire arrays and nanoporous structures, enhance light absorption by extending the propagation path of light. In nanowire array films, incident light undergoes multiple reflections and scattering events, increasing its interaction with the material and improving absorption efficiency. This enhanced light-harvesting capability significantly boosts the photoelectric conversion efficiency of photovoltaic devices.

#### *1.4.2.2 Optimization of charge transfer efficiency*

The nanoscale ordered structure enhances charge transport. In self-assembled nanochannels or layered structures, charge is efficiently transported in a specific direction. For example, in organic-inorganic hybrid photovoltaic films, self-assembled inorganic nanosheets provide fast electron transport, while the organic layer transports holes. This ordered charge transfer reduces recombination and improves charge collection efficiency, boosting photovoltaic device performance. Liu et al. [21] report a hybrid SAM strategy, in which the widely used SAM, Me-4PACz, is assembled with a multiple carboxylic acid functionalized aromatic molecule, 4,4′,4″-nitrilotribenzoic acid (NA), to form mixed SAMs (denoted as NA–Me) on the surface of nickel oxide (NiO) to fabricate inverted PSCs. The inclusion of NA molecules in the Me-4PACz layer improves the wettability of the perovskite solution on the HSL, reducing nanovoids and relieving interface stress. The interaction between Me-4PACz and NA's triphenylamine moiety reduces agglomeration, homogenizing Me-4PACz distribution, which enhances carrier extraction and reduces nonradiative recombination at the NiO–perovskite interface (**Figure 9**). The inverted devices based on formamidinium-cesium perovskite with the composition of $FA_{0.95}Cs_{0.05}PbI_3$ achieved optimal PCEs of 26.69% with a high open-circuit voltage ($V_{OC}$) of 1.201 V (certified steady-state efficiency of 26.54%) and 23.06% (certified 22.74%) for small-area PSCs.

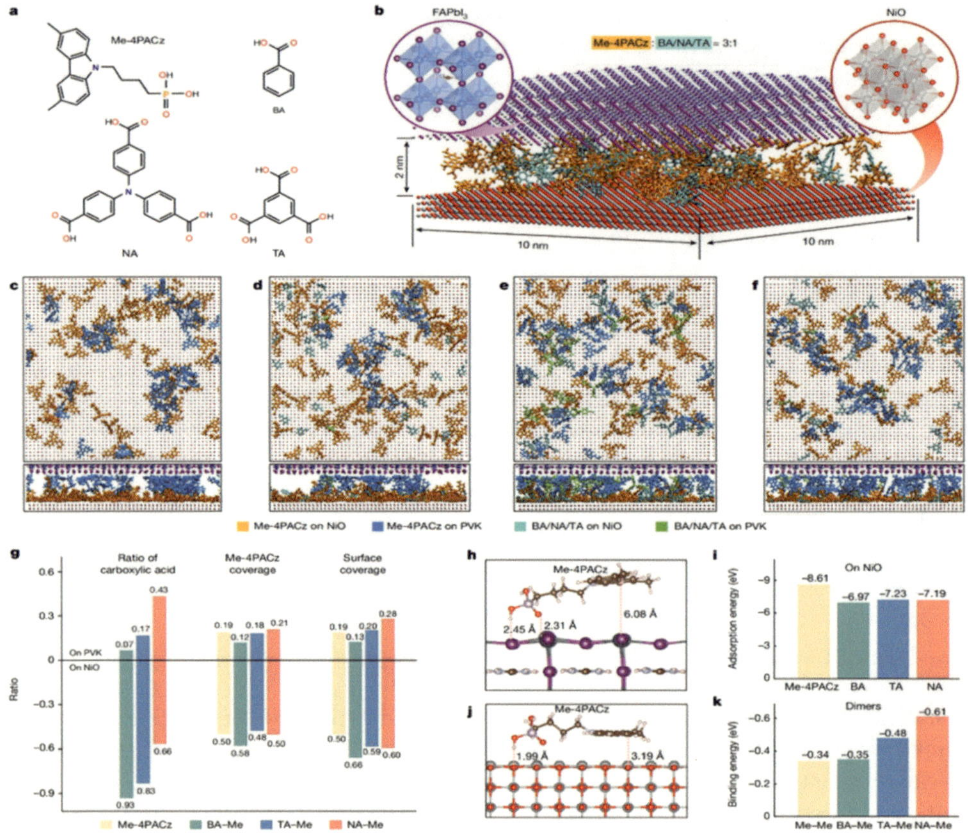


**Figure 9.**
*(a) Chemical structure of Me-4PACz and BA, NA, and TA. (b) Illustration of the perovskite/SAMs/NiO heterojunction model for molecular dynamics simulations. (c–f) Top and side views of molecular representations of the heterojunctions with Me-4PACz, BA–Me, NA–Me, and TA–Me hybrids. Me-4PACz is shown in orange and blue for adsorption on NiO and perovskite. Carboxylic acid molecules are in cyan and green for adsorption on NiO and perovskite. Ni and O atoms are in gray and red. Perovskite is omitted in the top view, with I atoms highlighted in purple in the side view. (g) Adsorption analysis of the perovskite and NiO surface in the heterojunction models. (h) The optimized adsorption structures of Me-4PACz at the NiO surface. (i) Adsorption energies for SAMs at the NiO surface. (j) Optimized adsorption structures of Me-4PACz on the perovskite surface. (k) Binding energies for the typical dimers of SAMs observed in the molecular dynamic simulations. Me–Me is the dimer of Me-4PACz [21]. Copyright (2024), The Author(s), under exclusive license to Springer Nature Limited.*

### 1.4.2.3 Breakthrough improvement of photoelectric conversion efficiency

Self-assembly technology enhances open-circuit voltage and short-circuit current density while reducing charge recombination, significantly improving the photoelectric conversion efficiency of thin-film photovoltaic devices. By controlling the self-assembly process and optimizing the nanoscale structure, the performance of many thin-film photovoltaic devices has been greatly enhanced. SAMs, with their solution processability and atomic-level structural tunability, are particularly effective for improving PSC performance and have been increasingly applied to PSC interfaces in the last 6 years (**Figure 10**) [22]. SAMs enhance thin-film perovskite solar cells by improving the perovskite layer's morphology, crystallinity, and reducing defects. SAMs fine-tune the polarity, controlling the interfacial dipole moment, work function, and energy barrier for better charge extraction. They also isolate the perovskite layer from metallic contacts, reducing

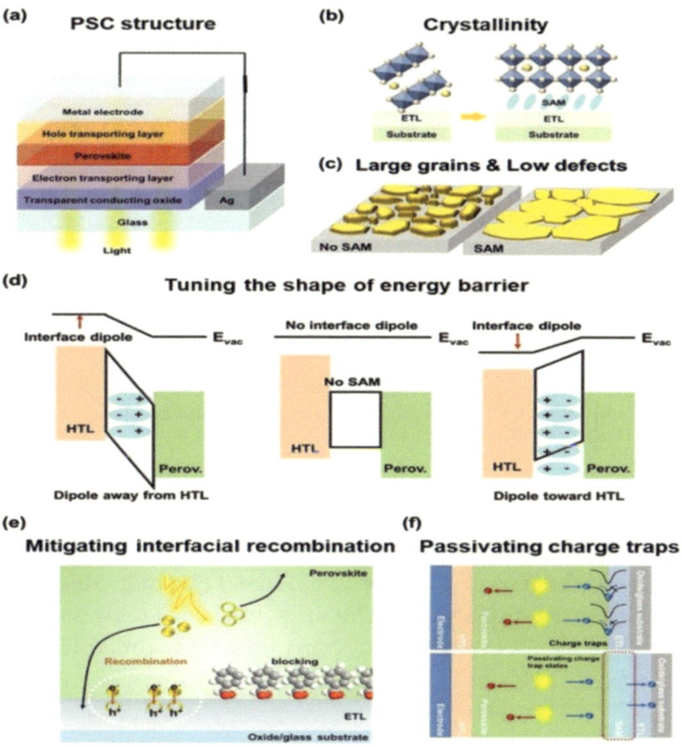


**Figure 10.**
*(a) Typical structure of conventional PSCs (n-i-p structure). (b) Schematic representations that summarize the effects of SAM-based interfacial engineering on crystallinity. (c) crystal morphology and defects. (d) energy barrier. (e) interfacial charge recombination. (f) charge traps [22]. Copyright (2020), Wiley-VCH GmbH.*

charge recombination, device hysteresis, and passivating charge trap states. With these improvements, SAMs have helped push the photoelectric conversion efficiency of organic-inorganic hybrid perovskite solar cells beyond 26%, highlighting the potential of self-assembly in advancing thin-film photovoltaics.

The self-assembly structure can enhance the open-circuit voltage of photovoltaic devices by optimizing the material's energy level structure. In quantum dot solar cells, self-assembly allows precise control over the spacing and interaction between quantum dots, reducing nonradiative charge recombination at the interface and improving the open-circuit voltage. Liao et al. [23] compared the behavior of MPA-BT-CA, MPA-BT-BA, and MPA-BT-RA grown on ITO based on self-assembly hole contacts, using DFT calculations to simulate the optimized stacking model, stacking each SAM in three layers. As shown in **Figure 11a**, MPA-BT-CA can maintain A tighter intermolecular stack (3.50 A) than MPA-BT-BA (3.73 A). At the same time, MPA-BT-RA shows an inclined stacking mode, which is attributed to the rotational degree of freedom hybrid carbon atom (-$CH_2$-) of $sp^3$ on the RA unit, which is attached with a -COOH group. The stacking of MPA-BT-RA molecules does not exhibit significant π-π stacking except for multiple O···H hydrogen bonds, which should lead to irregular arrangement during their self-assembly. At the same time, multiple hydrogen bonds can also induce strong aggregation of MPA-BT-RA in the thin film state (**Figure 11b**). Compared to the bare ITO glass (4.40 eV), the work function (WF) increases to 4.66 eV for MPA-BT-BA-modified ITO and further to 4.79 eV for

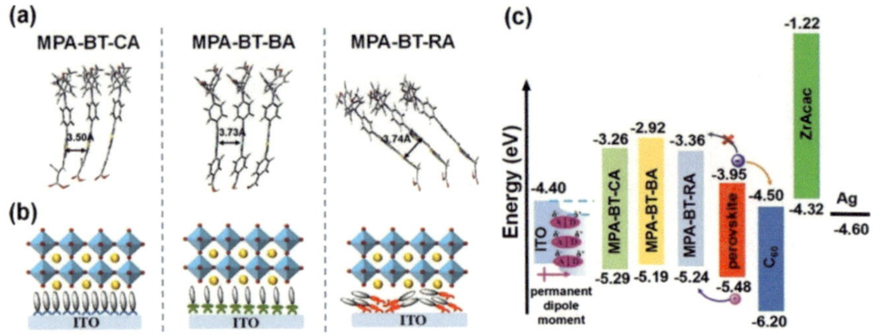


**Figure 11.**
*Optimized stacking models based on DFT calculations. (b) Schematic diagram of possible molecular organization behavior in the self-assembled films for three SAMs. (c) Schematic illustration of the device energy level diagram [23]. Copyright (2021), ELSEVIER B.V. and Science Press.*

MPA-BT-CA-modified ITO, improving energy alignment with the perovskite valence band (−5.48 eV) and enhancing hole transfer and $V_{OC}$. However, the MPA-BT-RA-modified ITO shows a decreased WF of 4.12 eV, likely due to its thick self-assembled layer. The HOMO energy levels of the SAMs align well with perovskite, supporting a high $V_{OC}$ and efficient electron extraction, transportation, and hole-blocking.

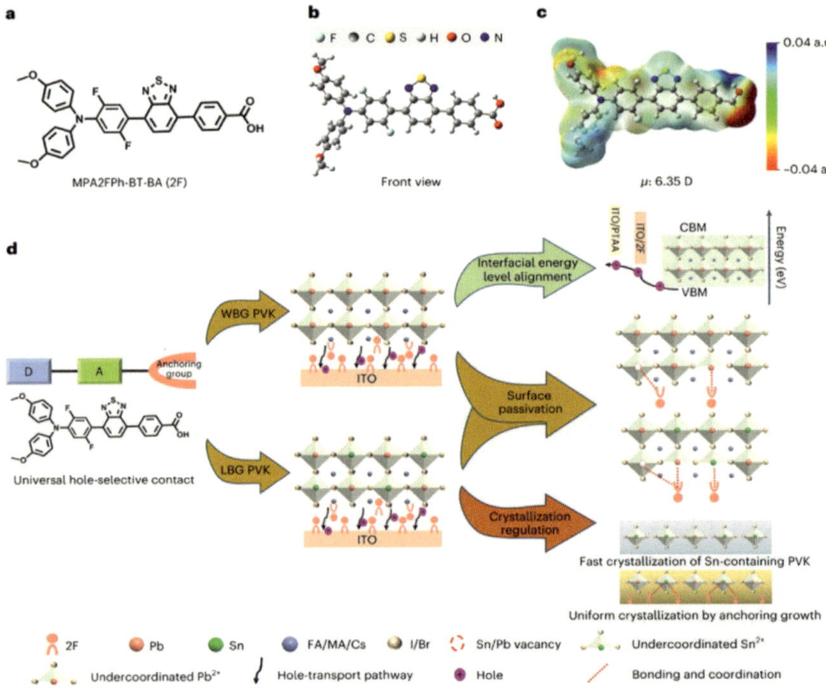


**Figure 12.**
*(a-c), Molecular structure, front view, and electrostatic surface potential and dipole moment of 2F, the corresponding dipole moment (μ) is 6.35 D, where D is Debye. Color bar, electrostatic surface potential. (d) Interconnection mechanisms between ITO, 2F, and perovskites. PVK, perovskite; VBM, valence band maximum; CBM, conduction band minimum; D, donor; A, acceptor [24]. Copyright (2023), The Author(s), under exclusive license to Springer Nature Limited.*

Self-assembly can effectively increase short-circuit current density by enhancing light absorption and optimizing charge transport. In dye-sensitized solar cells, self-assembly forms an ordered dye molecular layer on nanoporous titanium dioxide electrodes, increasing the light absorption cross-section and improving photogenerated carrier injection, thereby boosting short-circuit current density. This approach offers a promising method for achieving efficient and stable photovoltaic devices, contributing significantly to the development of renewable energy. Zhu et al. [24] designed a donor–acceptor-type molecule MPA2FPh-BT-BA (denoted as 2F) to replace PTAA and PEDOT:PSS as an efficient hole-selective contact employed in different PSCs with wide-bandgap (WBG) (1.77 eV $FA_{0.8}Cs_{0.2}Pb(I_{0.6}Br_{0.4})_3$ and 1.68 eV $FA_{0.8}Cs_{0.2}Pb(I_{0.8}Br_{0.2})_3$), low-bandgap (LBG) (1.25 eV $FA_{0.6}MA_{0.3}Cs_{0.1}Pb_{0.5}Sn_{0.5}I_3$) and normal-bandgap (NBG) (1.57 eV $FA_{0.8}Cs_{0.2}Pb(I_{0.95}Br_{0.05})_3$) perovskite absorbers. Self-assembly can effectively increase short-circuit current density by enhancing light absorption and optimizing charge transport. In dye-sensitized solar cells, self-assembly forms an ordered dye molecular layer on nanoporous titanium dioxide electrodes, increasing the light absorption cross-section and improving photogenerated carrier injection, thereby boosting short-circuit current density. This approach offers a promising method for achieving efficient and stable photovoltaic devices, contributing significantly to the development of renewable energy (**Figure 12**). As a result, their best-performing 1.77 eV WBG and 1.25 eV LBG devices achieve champion PCEs of 19.33% (certified 19.09%) and 23.24%, respectively, enabling a remarkably improved PCE (up to 27.22%, certified 26.3%) of 2F-tailored all-perovskite tandems with enhanced operational stability.

## 2. Self-assembly mechanisms and design principles

### 2.1 The underlying mechanisms that drive self-assembly

Self-assembly (SA) is the ultimate dream of the lazy scientist: simply mix these objects and the forces of nature will assemble them into the desired structure. Over the past few decades, the appeal of this vision has inspired countless chemists, physicists, and engineers to build self-assembling systems from molecules to macroscopic scales. In molecular science, SA provides the basis for the crystallization of organic [25, 26] and inorganic [27] molecules, and is at the heart of supramolecular chemistry, of which [28], the "instructions" for how to assemble larger entities. In nanotechnology, SA is the foundation for various types of molecular structures (e.g., Langmuir-Blodgett films, self-assembled monolayers [29], and amphiphilic fibers [30]) as well as higher-order structures (nanotubes [31]/or nanorods [32]) constructed from nanoparticles. At the microscopic scale, there are currently self-assembled fluid machines [33] and micromixers [34], complex microparticles built from prepolymer patches, and artificial opals [35] composed of millions of colloidal particles. Some of the macroscopic manifestations of SA include 2D and 3D structures driven by capillary forces, self-assembled electronic circuits [36, 37], and unusual crystals of charged components [38, 39]. Some SA strategies have found industrial applications. For example, Alien Technologies has developed a fluid self-assembly method to fabricate radio frequency identification (RFID) tags using shape recognition and fluid transport at the microscopic scale [40].

Self-assembly (SA) often relies on fine-tuning interparticle forces, which can be easier than manipulating components with optical tweezers. However, programming correct interactions in self-assembled molecules is typically based on intuition rather than predictive formulas. Understanding the underlying mechanisms through reverse engineering is crucial for advancing SA. While SA has demonstrated impressive potential, we are far from achieving the flexibility and creativity found in nature. In cells and organisms, dynamic self-assembly builds functional aggregates outside thermodynamic equilibrium, such as DNA replication machinery and cytoskeletal fibers. Exploring these dynamic systems could lead to future adaptive, self-replicating, or self-healing materials.

### *2.1.1 Chemical bonds: Self-assembled atomic-scale "viscose" Van der Waals*

The key driving force in molecular self-assembly is chemical bonding. Nanocrystal self-assembly relies on the properties of individual nanocrystals, including morphology, surface ligands, and solvent dispersion. Noncovalent forces like van der Waals, electrostatic, and hydrogen bonds guide the nanocrystals into their thermodynamically stable state. Therefore, effective control of the balance and combination of these forces is beneficial to facilitate the self-assembly process and obtain the desired structure.

Van der Waals forces, essential for nanocrystal self-assembly, are intermolecular forces resulting from electromagnetic fluctuations between atoms, molecules, and host materials. These short-range, attractive forces are unsaturated and directional, with strength inversely proportional to the sixth power of the intermolecular distance ($1/r^{(6)}$) [41].

The van der Waals force interaction can be divided into three types: (1) Dispersion force (London interaction), a weak attractive force generated by instantaneous dipoles when molecules are close, present in all molecules. Its strength depends on the molecule's deformation [42]. (2) Dipole-dipole attractive force (Keeson interaction, orientation force), where permanent dipoles in polar molecules create electrostatic attractions when aligned in opposite directions [43]. The strength of this force is linked to the molecule's dipole vector. Ji et al. reported that dipole-dipole attraction drives the formation of highly ordered $CsPbBr_3$ NWs, enhancing their stability [44]. (3) Inductive force (Debye interaction), which is the electrostatic attraction between an induced dipole and a permanent dipole in a molecule. The relationship between these forces depends on the polarity and deformability of the interacting molecules.

*Electrostatic interaction:* Like van der Waals forces, electrostatic interactions are crucial in nanocrystals but differ in that they depend on the charge distribution of the ligands. Electrostatic interactions can be attractive between oppositely charged nanocrystals and repulsive between equally charged ones. The repulsion prevents aggregation, while attraction promotes self-assembly in colloidal systems. By controlling nanocrystal concentration, ion size, and valence, self-assembly can be effectively achieved. Wang et al. [45] reported a new method for introducing gold bromide complexes into $CsPbBr_3$. The self-assembly process is caused by the electrostatic interaction between the gold bromide complex and the surfactant, as well as van der Waals forces between the carbon chains. Basically, the electrostatic interaction distance between two charges (r), the force can be considered $F = k_e q_1 q_2/r^2$ using Coulomb's law, where $q_1$ and $q_2$ are the amounts of the two charges ($ke = 8.988 \times 109\ N{\cdot}m^2{\cdot}C^{-2}$). However, in colloidal systems, the magnitude of the electrostatic force is more complex because the charge distribution in nanocrystals is sensitive to direction, interaction distance, solvent concentration, and polarity.

*Hydrogen bond*: Hydrogen bonds are special electrostatic dipole-dipole interactions between hydrogen donors and receivers, systems, and bond geometries. Hydrogen bonds are weaker than other driving forces, but they can still self-assemble between nanocrystals, which allows them to assemble into tightly packed superlattices [46].

### *2.1.2 Molecular orientation: A self-assembled oriented "navigator"*

Self-assembly relies on molecular recognition, where molecules interact through noncovalent bonds to form ordered aggregates. The synergy of weak interactions between molecules provides the driving force, maintaining structural stability and integrity. To assemble, molecules must complement each other in size and orientation, allowing for molecular rearrangement and precise spatial alignment.

*Spatial Arrangement Logic Dominated by Molecular Shape*: In molecular self-assembly, the interaction of structure and computation is key. Biological cells use structural rearrangements, like in the cytoskeleton, to integrate mechanical and chemical signals, though how information is processed remains unclear. Neural computing, such as the Hopfield model, shares hallmarks like decision-making, pattern recognition, and noise robustness, which are now being linked to molecular self-assembly. Multi-variant self-assembly, where components form different structures, reveals connections to neural models, suggesting collective dynamics at play even when individual molecules do not mimic nerve cell behavior. This dynamic process, driven by nucleation and growth, allows for diverse structural formations (**Figure 13**).

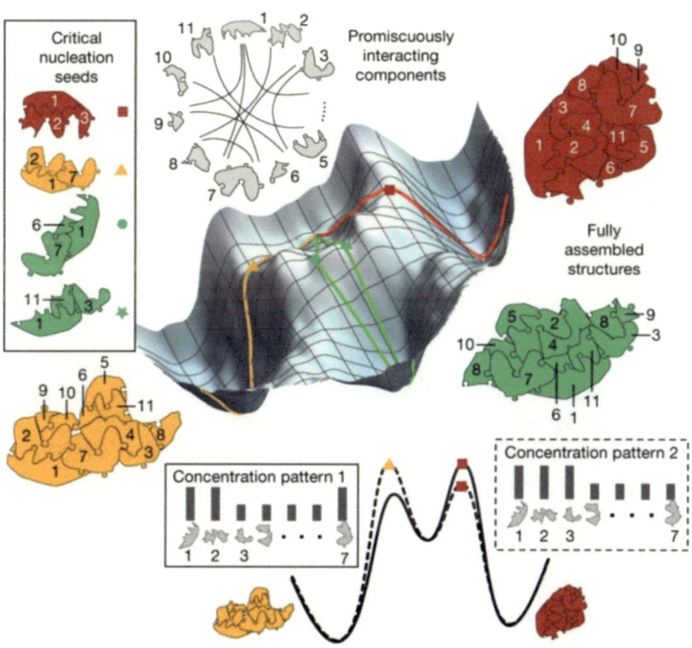


**Figure 13.**
*The nucleation pattern recognition framework suggests that when molecules can assemble into various structures, the nucleation process is influenced by high-dimensional concentration patterns. Assembly pathways, represented on an energy landscape, involve moving from the unassembled component basin through nucleation seeds (barriers) to the basin of the final structure. Co-localizing seeds of a high-concentration component reduces nucleation barriers, facilitating specific assembly pathways [47].*

### 2.1.3 Surface energy minimization

Nonequilibrium transport, interfaces, and mixing are ubiquitous in nature, at astrophysical and molecular scales, and in the range of high and low energy densities [48]. Processes like thermonuclear flashes, interstellar medium trajectories near black holes, hot spot formation in fusion, particle field interactions, mass ejections in solar flares, plasma instability, deep-sea convection, pollutant diffusion, plasma thrusters, nanofabrication, fossil fuel extraction, and premixed combustion are all driven by nonequilibrium interface dynamics [49]. These real-world environments are often characterized by sharply and rapidly changing flow fields and tiny dissipative and diffusion effects, which result in the formation of discontinuities (called frontiers or interfaces) between flow inhomogeneities (phases) on a macroscopic (continuous) scale.

The simplest way to determine wetting properties is by measuring the contact angle of droplets on a surface. The contact angle is influenced by surface roughness and chemical heterogeneity, and equations like the Young, Wenzel, and Cassie-Baxter models are commonly used for analysis. However, these models have limitations and may introduce errors when applied to surfaces with large contact angle hysteresis or those that do not meet the ideal conditions. The Young equation, which assumes an ideal, flat, chemically homogeneous, and non-reactive surface, is often used to quantify surface wettability:

$$\gamma_{sl} + \gamma_{\cos\theta} = \gamma_{so} \tag{1}$$

where $\gamma$, $\gamma_{sl}$, and $\gamma_{so}$ are liquid/gas surface tension, solid/liquid interfacial energy, and solid/gas surface energy, respectively. The apparent contact angle ($\theta$) is an equilibrium contact angle ($\theta Y$). However, since not all the real surfaces are ideal, models were developed to describe the contact angles on the real surfaces. The Wenzel and Cassie-Baxter models describe contact angles on real surfaces. The Wenzel model accounts for surface roughness but assumes chemical uniformity, while the Cassie-Baxter model considers chemical heterogeneity on flat surfaces. In the Wenzel model, surface roughness (r) is defined as the ratio of the actual surface area to the projected area. The Wenzel equation can be written as:

$$\cos(\theta *) = r cos(\theta Y) \tag{2}$$

where $\theta^*$ is the apparent contact angle and $\theta$Y is the equilibrium contact angle from Young's equation on an ideal solid without roughness.

In the Cassie-Baxter model, $f_1$ and $f_2$ are the area fractions of solids and air droplets on a substrate. The Cassie-Baxter equation can be written as:

$$\cos(\theta *) = f_1 \cos(\theta Y) - f_2 \tag{3}$$

where $\theta^*$ is the apparent contact angle and $\theta$Y is the equilibrium contact angle on the solid.

The Wenzel model suggests that surface roughness amplifies the wettability of the original surface, making hydrophilic surfaces more hydrophilic and hydrophobic surfaces more hydrophobic. The Cassie-Baxter model emphasizes the importance of the air area fraction under a droplet, with a larger air fraction leading to a higher contact angle. Both models, proposed decades ago, are now widely used and studied for superhydrophobic surfaces [50].

## 2.2 Self-assembly design principles for key links of photovoltaic devices

SAMs employed in perovskite solar cells (PSCs) have made breakthroughs in device efficiency and stability in single-junction and tandem configurations due to their unique and versatile ability to manipulate chemical and physical interface properties. The mechanical role of different SAMs in energy level regulation, interface modification, defect passivation, and charge transport. This is achieved by understanding how to fine-tune interfacial molecular interactions to mitigate charge recombination losses in reverse PSCs.

### *2.2.1 Self-assembly of light management dimensions*

In perovskite solar cells, self-assembly in light management enhances battery performance by forming ordered structures on the substrate. These structures guide light propagation in the perovskite layer, improving light absorption. Self-assembly of specific SAM molecules increases the effective absorption length of light, reduces transmission losses, and boosts short-circuit current density ($J_{SC}$), thereby improving power conversion efficiency (PCE). In light scattering and refraction regulation, the molecular structure of SAMs significantly impacts light behavior at media interfaces. Proper SAMs reduce reflection loss by stabilizing the interface through optimal anchoring group-substrate interaction. Adjusting spacer and terminal functional groups alters the refractive index, enhancing light scattering and promoting multiple reflections and absorption in the perovskite layer. This maximizes light energy utilization, improves light absorption efficiency, and boosts overall battery performance.

Designing better SAM molecules for light management in self-assembly involves several key considerations: (a) The SAM molecules should have specific functional groups and shapes to self-assemble into an ordered light capture structure on the substrate. (b) The self-assembly process should be precisely controlled by adjusting deposition conditions (temperature, concentration, time) to optimize light capture and absorption in the perovskite layer, improving efficiency. For light scattering and refraction regulation, molecular details must be considered. Strong anchoring groups are selected to stabilize the interface and reduce reflection, while spacer groups are adjusted to optimize the refractive index distribution. Terminal functional groups are designed for light response and compatibility with perovskites to enhance scattering and promote multiple light reflections. Using molecular simulations and experiments, SAM molecule structures are optimized to improve perovskite solar cell performance.

### *2.2.2 Self-assembly of charge transport layer*

At the interface of the charge transport layer, the design of the self-assembled layer follows the following important principles:

*Energy level matching principle:* The energy level structure of SAMs should align with the adjacent charge transport and perovskite layers. For SAMs at the hole transport layer (HTL) interface, the HOMO energy level must be close to the valence band of the perovskite for efficient hole extraction and transport while blocking electron reverse transport and reducing charge recombination. In efficient trans-perovskite solar cells, specific SAM molecules optimize the HOMO energy level through the design of functional groups, improving hole transport efficiency and reducing energy loss.

*Defect passivation principle:* SAMs should effectively passivate defects at the interface between the charge transport and perovskite layers. At the ETL-perovskite interface, SAMs interact with defect sites through active groups in their molecules, reducing defect state density. For instance, SAMs with phosphate or carboxyl groups can bind to uncoordinated metal ions or halogen vacancies on the perovskite surface, inhibiting charge recombination, improving Voc and FF, and enhancing overall battery performance.

*Stability Enhancement Principle:* The designed SAMs should enhance the interfacial stability between the charge transport and perovskite layers. Under prolonged lighting, temperature changes, and humidity, SAMs must maintain structural and property stability to prevent degradation or chemical reactions. SAMs with good chemical stability can ensure stable bonding with the charge transport and perovskite layers, promoting long-term device operation and extending the service life of the PV cell.

*Principles of molecular structure and assembly controllability:* The molecular structure of SAMs should be customizable to achieve an ideal self-assembly process and interfacial properties. By modifying components like anchoring groups, spacer groups, and terminal functional groups, the adsorption mode and molecular arrangement on the substrate can be controlled. For example, altering the type and chemical activity of anchoring groups influences the bonding strength and stability of SAMs, optimizing charge transport. Similarly, designing spacer group length and chemical structure can enhance the interaction between molecules and charge transport paths, improving interfacial charge transport efficiency.

## Acronym

| | |
|---|---|
| SAM | self-assembly materials |
| PSC | Perovskite solar cell |
| PCE | power conversion efficiency |
| FT-IR | Fourier transform infrared |
| FF | fill factor |
| VOC | open-circuit voltage |
| JSC | short-circuit current density |
| MPPT | maximum power-point tracking |
| TEM | transmission electron microscopy |
| NWs | nanowires |
| SA | self-assembly |
| HTL | hole transport layer |
| ETL | electron transport layer |
| FTO | fluorine-doped tin oxide |
| ITO | indium tin oxide |
| NiO | nickel oxide |
| WF | work function |
| PVD | physical vapor deposition |
| SEM | scanning electron microscopy |
| AFM | atomic force microscopy |
| XRD | X-ray diffraction |

## Author details

Zhenyu Zhang[1], Zhu Ma[2*], Zhuowei Du[2], Yi Chen[2], Qian Zhang[2], Hao Du[2], Zhuo Lv[2], Siming Huang[1], Maryam Shahrostami[3], Javad Maleki[3], Narges Yaghoobi Nia[4], Mojtaba Abdi-Jalebi[1*] and Mahmoud Zendehdel[4,5,6*]

1 Institute for Materials Discovery University College London Malet Place, London, UK

2 School of New Energy and Materials, Southwest Petroleum University, Chengdu, China

3 Department of Electrical and Computer Engineering, Tarbiat Modares University (TMU), Tehran, Iran

4 Kimia Solar Research Institute (K.S.R.I), Kimia Solar Company, Kashan, Iran

5 Iritaly Trading Company S.R.L., Via Volturno, Rome, Italy

6 Department of Electronics Engineering, University of Rome Tor Vergata, Roma, Italy

*Address all correspondence to: zhuma@swpu.edu.cn; m.jalebi@ucl.ac.uk and m.zendehdel@iritalytrading.com

## References

[1] Guha S. Thin-film photovoltaics: Buffer against degradation. Nature Energy. 2017;**2**:17057. DOI: 10.1038/nenergy.2017.57

[2] Li M, Liu M, Qi F, Lin FR, Jen AKY. Self-assembled monolayers for interfacial engineering in solution-processed thin-film electronic devices: Design, fabrication, and applications. Chemical Reviews. 2024;**124**:2138-2204. DOI: 10.1021/acs.chemrev.3c00396

[3] Li C et al. Self-assembly of block copolymers towards mesoporous materials for energy storage and conversion systems. Chemical Society Reviews. 2020;**49**:4681-4736. DOI: 10.1039/d0cs00021c

[4] Abelson A et al. Collective Topo-epitaxy in the self-assembly of a 3D quantum dot superlattice. Nature Materials. 2019;**19**:49-55. DOI: 10.1038/s41563-019-0485-2

[5] Liang Z et al. Fabrication of tunable aluminum nanodisk arrays via a self-assembly nanoparticle template method and their applications for performance enhancement in organic photovoltaics. Journal of Materials Chemistry A. 2018;**6**:3649-3658. DOI: 10.1039/c7ta10017e

[6] Wolff CM et al. Perfluorinated self-assembled monolayers enhance the stability and efficiency of inverted perovskite solar cells. ACS Nano. 2020;**14**:1445-1456. DOI: 10.1021/acsnano.9b03268

[7] Bi H et al. Selective contact self-assembled molecules for high-performance perovskite solar cells. eScience. 2024;**5**:100329. DOI: 10.1016/j.esci.2024.100329

[8] Li Y et al. Self-assembly of chiroptical ionic co-crystals from silver nanoclusters and organic macrocycles. Nature Chemistry. 2025;**17**:169-176. DOI: 10.1038/s41557-024-01696-6

[9] Zuo L et al. Enhanced photovoltaic performance of CH3NH3PbI3 perovskite solar cells through interfacial engineering using self-assembling monolayer. Journal of the American Chemical Society. 2015;**137**:2674-2679. DOI: 10.1021/ja512518r

[10] Yaghoobi Nia N, Matteocci F, Cina L, Di Carlo A. High-efficiency perovskite solar cell based on poly(3-Hexylthiophene): Influence of molecular weight and mesoscopic scaffold layer. ChemSusChem. 2017;**10**:3854-3860. DOI: 10.1002/cssc.201700635

[11] Yaghoobi Nia N et al. Doping strategy for efficient and stable triple cation hybrid perovskite solar cells and module based on poly(3-hexylthiophene) hole transport layer. Small. 2019;**15**:1904399. DOI: 10.1002/smll.201904399

[12] Yang Q et al. Solvent-actuated self-assembly of amphiphilic hole-transporting polymer enables bottom-surface passivation of perovskite film for efficient photovoltaics. Advanced Energy Materials. 2021;**11**:2100493. DOI: 10.1002/aenm.202100493

[13] Zendehdel M, Habibi MH, Hashemzadeh-Esfarjani MS. High efficiency photocatalytic mineralization of C.I. reactive red 2 using transparent self-assembled nanopin-like Titania with high preferential (101) orientation on alumina thin film in aqueous environments. Separation Science and Technology. 2014;**49**:2951-2959. DOI: 10.1080/01496395.2014.945186

[14] Habibi MH, Zendehdel M. Fabrication and characterization of self-assembled multilayer nanostructure Titania with high preferential (101) orientation on alumina thin films by layerby-layer dip-coating method. Current Nanoscience. 2010;**6**:642-647. DOI: 10.2174/157341310793348614

[15] Guo C et al. Light-induced Quinone conformation of polymer donors toward 19.9% efficiency organic solar cells. Energy & Environmental Science. 2024;**17**:2492-2499. DOI: 10.1039/d4ee00605d

[16] Tong X et al. Large orientation angle buried substrate enables efficient flexible perovskite solar cells and modules. Advanced Materials. 2024;**36**:2407032. DOI: 10.1002/adma.202407032

[17] He D et al. Decoupling nucleation and growth of crystal for DMSO-free self-assembly tin perovskite films and their optoelectronic application. Advanced Functional Materials. 2025;**19**:2420593. DOI: 10.1002/adfm.202420593

[18] Li Y et al. Interfacial self-assembly nanostructures: Constructions and applications. Small. 2024;**20**:2405318. DOI: 10.1002/smll.202405318

[19] Guo W et al. Self-assembled molecules fostering ordered spatial heterogeneity for efficient Ruddlesden-popper perovskite solar cells. Advanced Energy Materials. 2024;**14**:2401303. DOI: 10.1002/aenm.202401303

[20] Grzelczak M, Vermant J, Furst EM, Liz-Marzán LM. Directed self-assembly of nanoparticles. ACS Nano. 2010;**4**:3591-3605. DOI: 10.1021/nn100869j

[21] Liu S et al. Buried Interface molecular hybrid for inverted perovskite solar cells. Nature. 2024;**632**:536-542. DOI: 10.1038/s41586-024-07723-3

[22] Kim SY, Cho SJ, Byeon SE, He X, Yoon HJ. Self-assembled monolayers as Interface engineering nanomaterials in perovskite solar cells. Advanced Energy Materials. 2020;**10**:2002606. DOI: 10.1002/aenm.202002606

[23] Liao Q et al. Self-assembled donor-acceptor hole contacts for inverted perovskite solar cells with an efficiency approaching 22%: The impact of anchoring groups. Journal of Energy Chemistry. 2022;**68**:87-95. DOI: 10.1016/j.jechem.2021.11.001

[24] Zhu J et al. A donor–acceptor-type hole-selective contact reducing non-radiative recombination losses in both subcells towards efficient all-perovskite tandems. Nature Energy. 2023;**8**:714-724. DOI: 10.1038/s41560-023-01274-z

[25] Desiraju GR. Supramolecular synthons in crystal engineering—A new organic synthesis. Angewandte Chemie International Edition in English. 1995;**34**:2311-2327. DOI: 10.1002/anie.199523111

[26] Moulton B, Zaworotko MJ. From molecules to crystal engineering: Supramolecular isomerism and polymorphism in network solids. Chemical Reviews. 2001;**101**:1629-1658. DOI: 10.1021/cr9900432

[27] Aizenberg J, Muller DA, Grazul JL, Hamann DR. Direct fabrication of large micropatterned single crystals. Science. 2003;**299**:1205-1208. DOI: 10.1126/science.1079204

[28] Lehn J-M. Perspectives in supramolecular chemistry—From molecular recognition towards molecular information processing and self-organization. Angewandte Chemie International Edition. 1990;**29**:1304-1319. DOI: 10.1002/anie.199013041

[29] Whang D, Jin S, Wu Y, Lieber CM. Large-scale hierarchical organization of nanowire arrays for integrated nanosystems. Nano Letters. 2003;**3**:1255-1259. DOI: 10.1021/nl0345062

[30] Li M, Schnablegger H, Mann S. Coupled synthesis and self-assembly of nanoparticles to give structures with controlled organization. Nature. 1999;**402**:393-395. DOI: 10.1038/46509

[31] Park S-J, Lazarides AA, Mirkin CA, Letsinger RL. Directed assembly of periodic materials from protein and oligonucleotide-modified nanoparticle building blocks. Angewandte Chemie International Edition. 2001;**40**:2909-2912. DOI: 10.1002/1521-3773(20010803)40:15 <2909::AID-ANIE2909>3.0.CO;2-O

[32] Yan M, Zhang HT, Widjaja EJ, Chang RPH. Self-assembly of well-aligned gallium-doped zinc oxide Nanorods. Journal of Applied Physics. 2003;**94**:5240-5246. DOI: 10.1063/1.1608473

[33] Grzybowski BA, Radkowski M, Campbell CJ, Lee JN, Whitesides GM. Self-assembling fluidic machines. Applied Physics Letters. 2004;**84**:1798-1800. DOI: 10.1063/1.1664019

[34] Ottino JM, Wiggins SR, Campbell CJ, Grzybowski BA. Microfluidic mixers: From microfabricated to self-assembling devices. Philosophical Transactions of the Royal Society of London. Series A: Mathematical, Physical and Engineering Sciences. 2004;**362**:1069-1086. DOI: 10.1098/rsta.2003.1363

[35] Fialkowski M, Bitner A, Grzybowski BA. Self-assembly of polymeric microspheres of complex internal structures. Nature Materials. 2005;**4**:93-97. DOI: 10.1038/nmat1267

[36] Clark TD, Tien J, Duffy DC, Paul KE, Whitesides GM. Self-assembly of 10-μm-sized objects into ordered three-dimensional arrays. Journal of the American Chemical Society. 2001;**123**:7677-7682. DOI: 10.1021/ja010634l

[37] Clark TD, Ferrigno R, Tien J, Paul KE, Whitesides GM. Template-directed self-assembly of 10-μm-sized hexagonal plates. Journal of the American Chemical Society. 2002;**124**:5419-5426. DOI: 10.1021/ja020056o

[38] Gracias DH, Tien J, Breen TL, Hsu C, Whitesides GM. Forming electrical networks in three dimensions by self-assembly. Science. 2000;**289**:1170-1172. DOI: 10.1126/science.289.5482.1170

[39] Boncheva M et al. Magnetic self-assembly of three-dimensional surfaces from planar sheets. Proceedings of the National Academy of Sciences. 2005;**102**:3924-3929. DOI: 10.1073/pnas.0500807102

[40] Yeh HJJ, Smith JS. Fluidic self-assembly for the integration of Gaas light-emitting diodes on Si substrates. IEEE Photonics Technology Letters. 1994;**6**:706-708. DOI: 10.1109/68.300169

[41] Bishop KJM, Wilmer CE, Soh S, Grzybowski BA. Nanoscale forces and their uses in self-assembly. Small. 2009;**5**:1600-1630. DOI: 10.1002/smll.200900358

[42] Wagner JP, Schreiner PR. London dispersion in molecular chemistry—Reconsidering steric effects. Angewandte Chemie International Edition. 2015;**54**:12274-12296. DOI: 10.1002/anie.201503476

[43] Battezzati M, Magnasco V. Interaction energy of coupled

permanent dipoles in oriented molecules. Chemical Physics Letters. 2010;**496**:362-364. DOI: 10.1016/j.cplett.2010.07.064

[44] Ji Y, Wang M, Yang Z, Ji S, Qiu H. Nanowire-assisted self-assembly of one-dimensional nanocrystal Superlattice chains. Journal of Materials Chemistry C. 2019;7:8471-8476. DOI: 10.1039/C9TC02118C

[45] Luo D, Yan C, Wang T. Interparticle forces underlying nanoparticle self-assemblics. Small. 2015;**11**:5984-6008. DOI: 10.1002/smll.201501783

[46] Zendehdel M, Yaghoobi Nia N, Nasr-Esfahani M, Farahani P, Karbaschi MR. A combined computational and experimental study on the hydrogen bonding with chloride ion in a crab-claw like site of a new chromium Schiff Base complex. Inorganica Chimica Acta. 2016;**447**:150-161. DOI: 10.1016/j.ica.2016.03.040

[47] Evans CG, O'Brien J, Winfree E, Murugan A. Pattern recognition in the nucleation kinetics of non-equilibrium self-assembly. Nature. 2024;**625**:500-507. DOI: 10.1038/s41586-023-06890-z

[48] Abarzhi SI, Goddard WA. Interfaces and mixing: Nonequilibrium transport across the scales. Proceedings of the National Academy of Sciences of the United States of America. 2019;**116**:18171-18174. DOI: 10.1073/pnas.1818855116

[49] Bell JB, Day MS, Rendleman CA, Woosley SE, Zingale M. Direct numerical simulations of type Ia supernovae flames. II. The Rayleigh-Taylor instability. The Astrophysical Journal. 2004;**608**:883. DOI: 10.1086/420841

[50] Seo K, Kim M, Ahn JK, et al. Effects of drop size and measuring condition on static contact angle measurement on a superhydrophobic surface with goniometric technique. Korean Journal of Chemical Engineering. 2015;**32**:2394-2399. DOI: 10.1007/s11814-015-0034-x

Chapter 3

# Self-Assembly in Thin-Film Photovoltaics: Advancing Energy Conversion through Tailored Nanostructures

*Zhenyu Zhang, Zhu Ma, Zhuowei Du, Yi Chen, Qian Zhang, Hao Du, Zhuo Lv, Siming Huang, Maryam Shahrostami, Javad Maleki, Narges Yaghoobi Nia, Mahmoud Zendehdel and Mojtaba Abdi-Jalebi*

## Abstract

Self-assembled monolayers (SAMs) have emerged as pivotal molecular tools for interface engineering in thin-film photovoltaic technologies, including perovskite solar cells (PSCs), dye-sensitized solar cells (DSSCs), organic photovoltaics (OPVs), and quantum dot photovoltaics (QDPVs). By enabling precise control over energy-level alignment, defect passivation, crystallization kinetics, and interfacial stability, SAMs significantly enhance both the efficiency and long-term durability of photovoltaic devices. This chapter presents a comprehensive review of the roles and mechanisms of SAMs in various solar cell architectures, emphasizing their impact on charge transport, light harvesting, and mitigation of ionic migration. In PSCs, SAMs modulate interfacial dipoles, promote large-grained film growth, and suppress nonradiative recombination, contributing to power conversion efficiencies exceeding 26%. In DSSCs and OPVs, SAMs optimize electrode interfaces and enable controlled phase separation, while in QDPVs, self-assembly facilitates the formation of dense, ordered nanocrystal superlattices for enhanced carrier transport. We also explore recent advances in tandem photovoltaic devices, where SAMs enable conformal coverage on textured substrates and play a critical role in bridging wide- and narrow-bandgap subcells. This work highlights the interdisciplinary potential of molecular self-assembly as a scalable and tunable strategy to overcome persistent limitations in next-generation solar technologies.

**Keywords:** self-assembled monolayers (SAMs), interface engineering, perovskite solar cells (PSCs), dye-sensitized solar cells (DSSCs), organic photovoltaics (OPVs), quantum dot photovoltaics (QDPVs), tandem solar cells, molecular dipole modification, thin-film photovoltaics, energy-level alignment, ionic migration suppression, charge transport enhancement, light trapping and absorption, surface morphology control, defect passivation

## 1. Introduction

SAMs are gaining attention in the PSCs industry due to their ability to enhance device performance through effective interface engineering. A brief introduction to the PSCs and SAM-based PSCs is provided in the following sections.

### 1.1 Perovskite photovoltaics: An overview

PSCs have become popular due to their outstanding properties like high power conversion efficiency (PCE), tunable bandgap, low-cost and easy manufacturing process, and long carrier lifetime and mobilities [1]. These cells are composed of different layers, including transparent conductive oxides (TCOs) such as fluorine-doped tin oxide (FTO) or indium tin oxide (ITO) as the front electrode, an electron transport layer (ETL), an active layer which absorbs light (perovskite materials), a hole transport layer (HTL), and noble metals like gold as the back electrode from bottom to top provided in **Figure 1** [2].

### 1.2 Significance of self-assembly in PSCs

Although the PSCs have made significant progress since they were first introduced, with the efficiency enhancement from 3.8% in 2009 to over 26% lately, they are still suffering from low efficiency and instability [3]. The two mentioned drawbacks can be improved through the interface engineering of layers. In doing so, self-assembly has proven worthy as a remarkable approach to optimizing the interfacial properties at the molecular level. The interfacial properties can be controlled through employing self-assembling molecules to form SAMs [4, 5], which is essential in order to reach the following achievements: (1) enhancing device stability, the risk of degradation processes such as chemical interactions, moisture penetration, and ion migration is reduced by utilizing SAMs that serve as protective shields between layers. (2) Improving charge carrier extraction, charge carrier extraction is improved through making efficient energy-level alignment and limiting the interfacial traps to ease the charge transfer between the perovskite layer and the charge transport layers [6]. (3) Passivating defects, the perovskite defects lead to recombination centers, resulting in lower PCE, which is passivated efficiently by using SAMs through reducing nonradiative recombination losses [7, 8].

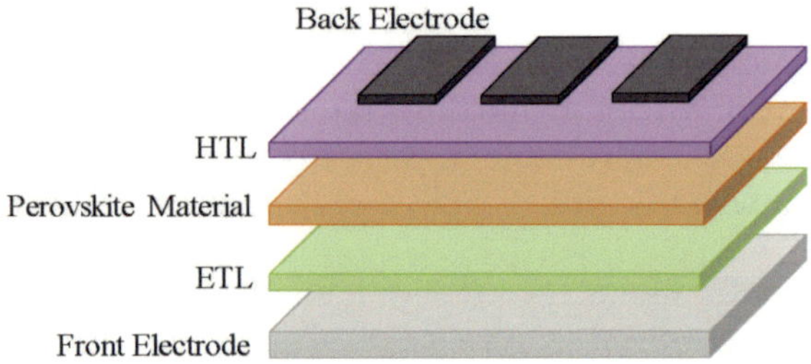


**Figure 1.**
*A common PSC structure.*

## 1.3 Fundamentals of self-assembly in PSCs

In this section, the principles of SAMs-based PSCs, the SAMs position in the device's structure, and their different types of depositions are discussed.

### *1.3.1 Principles of self-assembly in PSCs*

SAMs effectively improve the efficiency and stability of the PSCs through various methods such as morphological adjustment, energy-level modification, and interface engineering.

#### *1.3.1.1 Morphological adjustment*

SAMs consist of anchor, linker, and terminal groups (**Figure 2**). Anchor groups bond to substrates, passivating defects. Linker groups define spacer length and create barriers, while terminal groups modify interfacial properties, influencing energy levels, wettability, and film quality [3]. The linker group connects anchor and terminal groups, affecting electrical isolation and molecular packing. Its electronic properties influence charge transport, with conjugated backbones reducing attenuation.

#### *1.3.1.2 Energy-level modification*

Research indicates that SAMs can modify energy levels by creating dipole moments at interfaces, thus influencing work function (WF). They also passivate defects in metal oxides and perovskites through chemical bonding, enhancing electronic states and stability.

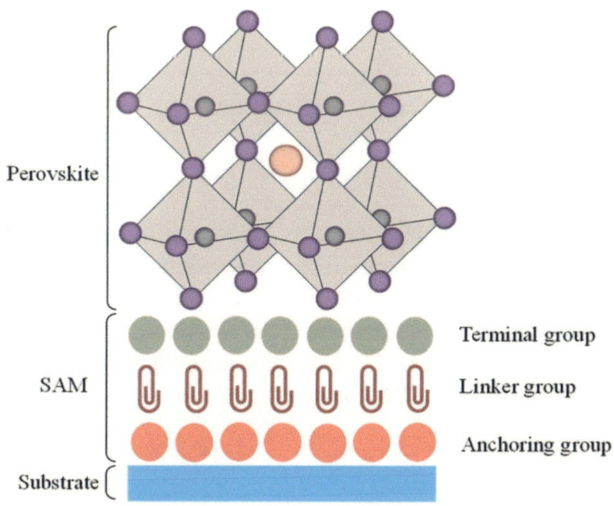


**Figure 2.**
*The SAMs composition in a common PSC.*

#### *1.3.1.3 Interface engineering*

Furthermore, SAMs can optimize the composition at the interface between metal oxides and perovskite layers, facilitating charge extraction and improving device efficiency. This simple yet effective interface engineering through SAM modification is a promising area for advancing highly efficient and stable PSC technology [9].

## 1.4 SAMs as ETL and HTL in PSCs

SAMs enhance the interactions between layers in perovskite solar cells. They can be applied on either side of the perovskite layer to facilitate charge transport to the hole or electron transport layers.

### *1.4.1 SAMs Interface between ETL and perovskite layer*

At the ETL-perovskite interface, SAMs enhance charge transport by modifying energy levels and reducing recombination losses. They achieve this by creating favorable dipole moments that optimize charge extraction, leading to improved device efficiency.

#### *1.4.1.1 Anchor group selection*

Common anchor groups, such as phosphates, carboxylic acids, and silanes, bond with metal oxide ETLs like $TiO_2$, $SnO_2$, and ZnO. These groups reduce the exposure of hydroxyl (-OH) groups on the metal oxide surface by forming anchoring bonds. This replacement leads to a more hydrophobic surface, improving the quality of the perovskite film [10].

#### *1.4.1.2 Terminal group selection*

Fullerene-functionalized SAMs are effective in enhancing electron transport between the perovskite and ETLs like $TiO_2$. The strong interaction between fullerene groups and perovskite iodide helps passivate defects at the interface, allowing electrons to move freely and minimizing recombination losses [11].

### *1.4.2 SAMs Interface between perovskite layer and HTL*

Similarly, at the perovskite-HTL interface, SAMs facilitate better interaction and stability. They passivate surface defects in the perovskite layer and adjust the WF of the HTL, which aids in effective charge transfer. This dual functionality of SAMs not only enhances crystallization of the perovskite material but also contributes to the overall stability and performance of perovskite solar cells, making them an essential component in modern photovoltaic technology.

#### *1.4.2.1 Anchor unit selection*

Anchor units, such as phosphonic acids, interact with the HTL to passivate defects and redistribute charges at the interface [12]. This alters the HTL's WF, enhancing hole transport efficiency.

#### *1.4.2.2 Terminal group selection*

The terminal groups of SAMs influence perovskite crystallization, reduce surface defects, and adjust energy levels. They also control dipole moments and electron distribution, impacting charge transfer and device performance. For example, modifying SAM molecules like phenylphosphonic acid (PPA) with different chemical groups can tune WF and dipole direction, optimizing charge extraction and reducing energy losses. This improves open-circuit voltage ($V_{OC}$), short-circuit current ($J_{SC}$), and fill factor (FF), leading to better overall performance [13–15].

### 1.5 SAMs deposition techniques in PSCs

SAMs can be deposited onto substrates using different techniques, including vapor processing, solution processing, thermal evaporation, and incorporation into perovskite solutions. (1) Vapor processing, vapor processing involves exposing the substrate to a saturated atmosphere of volatile SAM compounds, allowing the formation of a self-assembled layer. Annealing is then performed to strengthen the bond with the substrate and remove excess unbound molecules. (2) Solution processing, solution processing is the most common method for SAM deposition, utilizing techniques such as dip-coating, spin-coating, and spray-coating. (3) Dip-coating, the substrate is immersed in a SAM solution, allowing chemical adsorption. This method requires careful control of solvent, concentration, and immersion time, but can involve toxic solvents and lengthy immersion periods. (4) Spin-Coating, this technique involves applying SAM solutions at high speeds to achieve uniform layers without always needing solvent washing. (5) Spray-coating: This method requires precise control for high-quality layers. Air-brush coating offers a simpler alternative, where SAMs are sprayed onto substrates, dried with nitrogen, and then annealed. (6) Thermal evaporation, recent studies have introduced physical vapor deposition (PVD) *via* thermal evaporation for high-quality SAM hole transport layers in PSCs [16]. This approach matches the performance of solution-processed layers and is advantageous for commercial production due to its ability to create uniform coatings on textured surfaces, enhancing yield and reproducibility. (7) Incorporation into perovskite solutions, SAMs can also be incorporated directly into the perovskite precursor solution, forming a compact layer on metal oxide surfaces during film processing. This co-deposition simplifies manufacturing and helps passivate perovskite films, further improving device performance [13].

In all solution processing methods, excess SAMs are typically removed through rinsing or washing after thermal annealing to ensure strong covalent bonding.

### 1.6 Strategies to enhance light absorption

Enhancing the light absorption of the solar cell plays a crucial role in improving the total efficiency of the cell. In doing so, controlling the crystallinity and surface morphology of perovskite films are essential.

#### *1.6.1 Morphology control for enhanced light trapping*

The crystallinity and surface morphology of perovskite films are key to improving light trapping and overall solar cell efficiency. Large, high-quality crystal domains enable better charge separation and extraction, while small grains with numerous grain boundaries increase charge recombination.

Perovskite film growth involves nucleation followed by grain formation during drying and annealing [17, 18]. SAMs aid this process by reducing surface friction and promoting larger grain sizes with fewer defects. These improvements enhance light absorption and charge transport, contributing to better device performance. Techniques like scanning electron microscopy (SEM), atomic force microscopy (AFM), and X-ray diffraction (XRD) are used to analyze these morphological changes, highlighting the role of SAMs in optimizing perovskite structure for efficient light trapping.

Additionally, incorporating a SAM beneath the perovskite crystal film improves the compatibility between the perovskite and charge transport layers, promotes the growth of perovskite crystals with preferred orientation, and enhances crystal domain size and quality while minimizing defects [7].

#### *1.6.1.1 Antireflective coatings*

Antireflective coatings play an important role in enhancing light absorption in PSCs by minimizing reflection losses at the surface. The incorporation of SAMs as antireflective layers can significantly improve the optical properties of these devices. By tailoring the surface morphology and reducing the refractive index mismatch between layers, SAMs facilitate better light management, allowing more photons to enter the active layer. This increased light absorption leads to enhanced charge generation and improved overall efficiency of PSCs. Recent advancements in nanostructured coatings, such as honeycomb-patterned surfaces created through nanoimprinting, demonstrate further potential for optimizing light trapping while maintaining compatibility with the delicate nature of perovskite materials [19].

#### *1.6.1.2 Textured substrates*

Textured substrates enhance light absorption in PSCs by scattering incoming light, increasing its path length within the active layer, and improving photon capture. Techniques like copper-seeding or employing FTO substrates create these rough surfaces. To further optimize these textured surfaces, SAMs can be integrated. SAMs improve interfacial properties, facilitating better charge transport and minimizing surface energy mismatches. This synergistic combination of textured substrates and SAMs results in reduced reflection losses and significantly enhanced light absorption, leading to a remarkable boost in the overall efficiency of perovskite solar cells [20, 21].

## 1.7 Inhibiting ionic migration

Although the ion migration in the PSCs has been eliminating the stability of solar cells for years, the introduction of SAMs techniques has provided a remarkable impact on the new PSCs research.

### *1.7.1 Understanding ion migration in PSCs*

Ionic migration is a significant challenge in PSCs, affecting their stability and efficiency. This phenomenon involves the movement of ions, such as halides (e.g., $I^-$)

and organic cations (e.g., methylammonium, $MA^+$), within the perovskite material due to electric fields, heat, or external stimuli. It can lead to reversible effects like current-voltage hysteresis, where device performance varies based on scan direction, as well as irreversible degradation from chemical reactions between migrating ions and adjacent layers. In methylammonium lead iodide ($MAPbI_3$), for instance, the volatility of $MA^+$ and the mobility of $I^-$ contribute to defect formation and phase instability, reducing PCE. Strategies such as substituting $MA^+$ with formamidinium ($FA^+$) or cesium ($Cs^+$) have shown potential in minimizing ion migration and enhancing thermal stability [22].

*1.7.2 Ion migration mitigation in PSCs using SAMs*

In PSCs, ion migration poses a significant challenge to stability and efficiency. SAMs have emerged as a promising solution by passivating defects and modulating interfaces. They can suppress nonradiative recombination and inhibit ion migration, thereby enhancing device stability under stress conditions. By facilitating hole extraction and reducing interfacial defects, SAMs contribute to improved power conversion efficiencies and prolonged lifetimes in PSCs. This approach highlights the potential of SAMs in mitigating ion migration-related issues, offering a pathway to more robust and efficient solar devices [23].

*1.7.3 Advances in PSCs ion migration mitigation using SAMs*

*1.7.3.1 SAM-treated $SnO_2$ layers*

Tumen-Ulzii et al. identified that the current-voltage hysteresis and degradation in PSCs are caused by the localization of positive ions, which is influenced by hydroxyl groups present on the $SnO_2$ ETL. To address this issue, they applied a SAM treatment to deactivate these hydroxyl groups. This modification resulted in PSCs with no hysteresis and exceptional stability, showing no degradation even after 1000 hours of continuous illumination. These findings underscore the potential of SAMs in improving both the efficiency and durability of PSCs, advancing their prospects for commercialization [24].

*1.7.3.2 POZ-BT-PY SAMs*

Hung et al. synthesized innovative SAM molecules, including DPA-B-PY, CBZ-B-PY, and POZ-BT-PY, to address interfacial defects caused by ion migration in inverted PSCs. Using a two-step method, POZ-BT-PY SAMs were layered on top of a CbzNaph (CNph) SAM to form a robust CNph SAM@pseudo-planar monolayer. This design improved charge transfer, passivated deep defects, and promoted perovskite crystal growth, leading to a PCE of 24.45% and an FF of 82.63% for POZ-BT-PY-modified devices [25].

### 1.8 Improving stability

The stability improvement of PSCs, caused by different factors, led scientists and researchers to find the SAMs-based PSCs an interesting topic to investigate, in order to resolve the interfacial layer interactions.

### *1.8.1 PSCs' stability enhancement using SAMs*

The stability of PSCs remains a major hurdle for practical use, as environmental factors like oxygen, moisture, and UV light degrade the perovskite layers, impairing charge transport and reducing PCE. UV exposure, for example, induces $PbI_2$ formation at interfaces, further destabilizing the device [26]. Additionally, reactive dangling bonds on perovskite surfaces create charge traps, leading to recombination losses and diminished performance. To mitigate these issues, the introduction of SAMs at the interfaces of PSCs has emerged as an effective strategy. SAMs work by passivating surface defects and deactivating reactive dangling bonds, thereby reducing the impact of moisture and oxygen on the perovskite layer. By providing a protective barrier, SAMs help stabilize the crystal structure and improve charge transport properties. This not only enhances the operational durability of PSCs but also maintains their efficiency over extended periods, making them more viable for commercial applications [7].

### *1.8.2 Advances in PSCs stability enhancement using SAMs*

PSCs face significant stability challenges due to degradation caused by oxygen, water, and light exposure. For example, UV light can generate $PbI_2$ at the perovskite interface, while water and oxygen infiltration degrade the perovskite layer, reducing charge transport efficiency and PCE. Introducing SAMs as HTLs has proven effective in mitigating these issues. Examples of advances in PSCs stability enhancement using SAMs include:

#### *1.8.2.1 Phosphonic acid-based SAMs*

Magomedov et al. developed a phosphonic acid SAM (V1036) applied directly to ITO substrates, avoiding high-temperature processes and reducing parasitic absorption. Devices with V1036 showed remarkable stability, retaining over 95% of their initial efficiency after 1000 hours of stress testing under heat and illumination. Mixed SAMs, such as those incorporating butyl phosphonic acid, further enhanced interfacial coverage, reduced recombination losses, and provided stronger environmental protection [27].

#### *1.8.2.2 Fluorinated boronic acid SAMs*

Kara et al. demonstrated that fluorine-terminated SAMs on ITO significantly improved device stability by increasing ITO's WF and creating a protective barrier against degradation from PEDOT:PSS and environmental factors. Devices with 2F-SAM-modified cells retained 80% efficiency after 30 days, compared to only 25% for unmodified devices. These SAMs also optimized interfacial energy alignment, enhancing charge collection and reducing hysteresis [28].

#### *1.8.2.3 Thiol-functionalized SAMs*

Chih-Yu Chang et al. employed thiol-functionalized SAMs (e.g., MPTMS and MUTAB) on nickel oxide nanoparticles to improve the interface between the perovskite layer and electrodes. These SAMs enhanced adhesion and uniformity of ultrathin Ag films used as transparent electrodes, achieving a PCE of 16.2% for large-area devices (1.2 $cm^2$) with excellent ambient stability without encapsulation [29].

## 2. Applications in dye-sensitized solar cells (DSSCs)

### 2.1 The basic principles of dye-sensitized solar cells

#### *2.1.1 Background*

The Dye-sensitized solar cell (DSSC) is the third generation of solar cells that was first introduced by -O'Regan and Gratzel in 1991 [30, 31]. DSSCs consist of organic and inorganic materials, classified based on their light-absorbing layers. Inorganic dyes, like ruthenium, rhodium, and porphyrin, are synthetic metal complexes. Organic dyes can be either synthetic, such as D35 and D5, or derived from plant sources like leaves, fruit, and seeds, including anthocyanin, chlorophyll, and carotene [32]. In terms of stability and efficiency, the metal complex-based dyes are superior to the plant-extracted dyes, but the accessibility of the dye and environmental aspects of natural dyes are comparable with synthetic dyes. A dye like ruthenium synthesis requires a long time and a high cost due to its process in chromatography, but its electron-hole generation and stability are quite impressive; therefore, it is a highly efficient dye in DSSCs [33, 34]. The highest efficiency for ruthenium-based dye has been recorded to date is 14 and 14.2% [35], while natural dye-based solar cells exhibited a maximum efficiency of around 9.8% [32, 35]. The major constituents of DSSC are: (i) substrate; (ii) semiconductor; (iii) dye; (iv) electrolyte; and (v) counter electrode (CE). DSSC provides a sandwich-like appearance when the five different layers are sealed together. The performance of DSSC generally depends on the proper sealing of these layers as well as the types and nature of the major constituents used in it [36–38].

#### *2.1.2 Working mechanism of DSSC*

The DSSCs works on a very different basis from the first- and second-generation solar cells and is closely related to natural photosynthesis, where light absorption and charge carrier transport are carried out by different substances. Specifically, the components and basic processes of the DSSC are schematically shown in **Figure 3**. The basic process includes: (1) photoexcitation of the sensitizer to produce excitation dyes; (2) the band of metal oxide, Lead to the production of oxidized dyes; (3) electrons are transmitted to the anode and flow to the opposite electrode *via* the external circuit; (4) regeneration of oxidized dyes by receiving electrons from reduced redox pairs; (5) redox pair regeneration by accepting oxidized electrons from the anode at the cathode; (6) electronic recombination by providing electrons to the oxidized dyes; (7) electronic recombination by providing electrons to oxidized redox pairs; (8) the excited-state dye is relaxed to the ground state through a nonradiative decay process [39–47].

In DSSCs, processes (1)–(5) drive photoelectric conversion, while dark reactions (6) and (7) generate charge carriers that slightly impact efficiency due to their slower response. Additionally, electron, ion, and molecule migration in $TiO_2$ mesoporous and electrolyte layers influences performance. Optimization focuses on tuning the physicochemical properties of the four key components (**Figure 3**): (i) sensitized dyes, (ii) metal oxide semiconductor, (iii) redox pair electrolyte, and (iv) pair electrode. For high-efficiency solar cells, all components must be finely tuned. The core is a mesoporous oxide layer of nanometer-sized particles sintered on a conductive glass substrate (ITO or FTO). This layer supports the sensitizer, facilitates

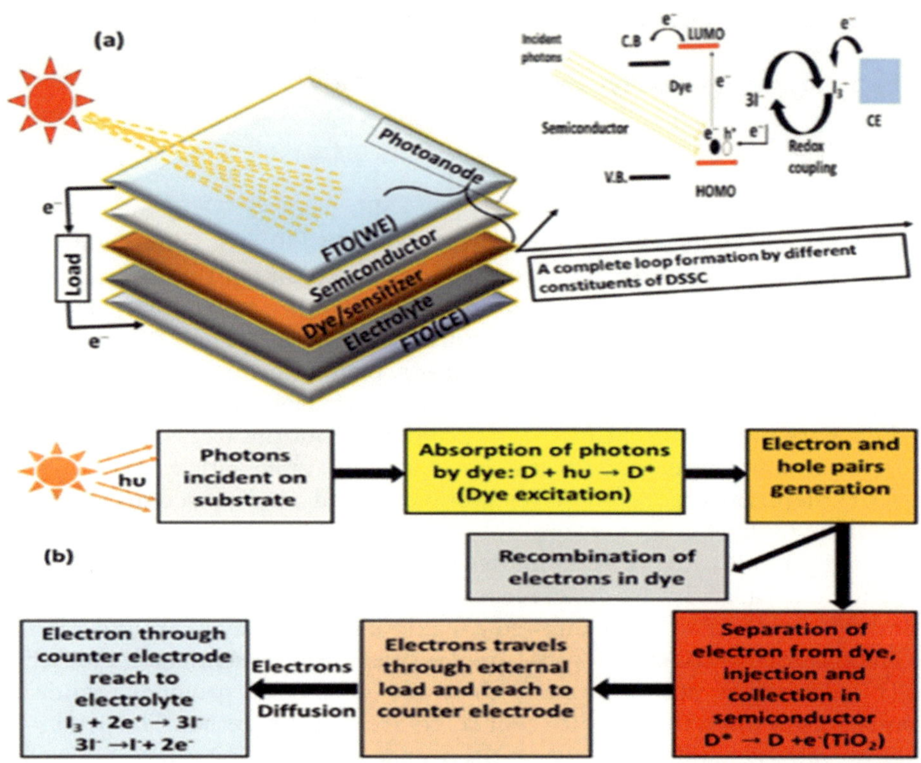


**Figure 3.**
*(a) Symbolic representation of the working principle and basic constituents of DSSC, (b) Schematic diagram with basic reactions involved in the working of DSSC [32]. Copyright (2024)* Elsevier Optical Materials.

electron transport, and enables redox diffusion. Titanium dioxide (anatase) is the most common DSSC semiconductor due to its stability, low cost, non-toxicity, and high bandgap (3.2 eV), enhancing the Fermi level and VOC [21, 22]. Indeed, various nanocomposite structures, including zinc tin oxide (ZTO) [48], copper-zinc oxide [49], polyoxometalates [50], zinc zirconate [51], zinc titanate [52], zinc ferrite [53], and CuO-ZnO [54], have also been explored as alternative working electrodes for high-performance DSSCs. Attached to the surface of the oxide semiconductor film is a monolayer of dye sensitizer. The role of the dye in DSSCs consists in acting as a molecular electron pump. As an effective dye sensitizer, it should have wide and strong absorption in the visible range, even in the NIR range, and strongly anchor groups (-COOH, -$H_2PO_3$, -$SO_3H$, etc.). Binding of the dye on the semiconductor surface, high stability and reversibility in the oxidation, ground and excited states, suitable redox potential associated with the semiconductor and charge mediator, and optical, thermal, and electrochemical stability [55]. The renewal of the highest efficiency of the DSSC is often accompanied by the discovery of new dyes.

## 2.2 Application of self-assembled monolayer membrane in dye-sensitized solar cells

SAMs in DSSCs enhance electrode materials for better performance and stability. Graphite-platinum self-assembled film counter electrodes offer high efficiency and low cost, making them a promising alternative. DSSCs consist of a $TiO_2$-based

*DOI: http://dx.doi.org/10.5772/intechopen.1012292*

photoanode, a redox electrolyte (e.g., $I/I_3$ and cobalt complex mediators), and a counter electrode (CE). Photoexcited dye molecules inject electrons into $TiO_2$, while $I_3$ is reduced at the CE, which catalyzes ion regeneration. Platinum, the conventional CE material, offers excellent conductivity and stability but is costly and inefficiently utilized. Reducing Pt usage with cost-effective manufacturing remains a key challenge for large-scale DSSC production.

### *2.2.1 Modify the opposite electrode to reduce the charge recombination*

Charge recombination significantly impacts DSSC performance. Self-assembled monolayers (SAMs) on electrodes reduce recombination and enhance efficiency by forming thin insulating layers. Long-chain alkyl SAMs create uniform insulation, while functional groups like carboxyl or amino groups improve surface energy and charge transport. Silica or gold nanoparticles are commonly used as substrates for azo dye self-assembly [56, 57]. A terminal thiol group usually binds to the dye structure, so the dye will anchor to the substrate [58]. In contrast, the photoisomerization of azo dyes on $TiO_2$ nanoparticles is much less studied than on the gold particles and azo dyes in solution [59]. Due to the development of DSSC, when gold nanoparticles are replaced by $TiO_2$ nanoparticles, many anchoring groups are available, such as carboxylates, phosphonates, salicylates, sulfonates, acetylpyruvate, dinaphthalene, nitro, phosphonate, pyridine, siloxane, and isohydroxamate groups. (**Figure 4**) [60] This provides a greater degree of freedom when designing future molecular switches than a single choice of sulfhydryl as an anchor for gold. Therefore, it would be interesting to explore this azo/$TiO_2$ nanoscale system from the perspective of intelligent surface design (photoinduced isomerization) and photoelectric (DSSC).

Carboxylic acids and phosphonic acids can form ordered SAM on metal or metal oxide surfaces [61, 62]. Tao et al. used alkanoic acids with various chain lengths to form the carboxylate SAM on noble metal surfaces and analyzed its structure by elliptical polarization and reflection-absorption infrared spectroscopy [63]. Breen et al.

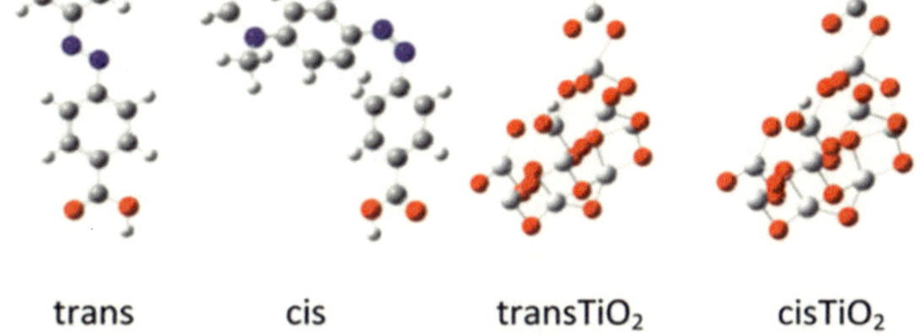


**Figure 4.**
*DFT-optimized structures for trans, cis, trans$TiO_2$, and cis $TiO_2$. Red, oxygen; blue, nitrogen; gray, carbon; small white, hydrogen; large white, titanium. Copyright (2014)* ACS Applied Materials & Interfaces.

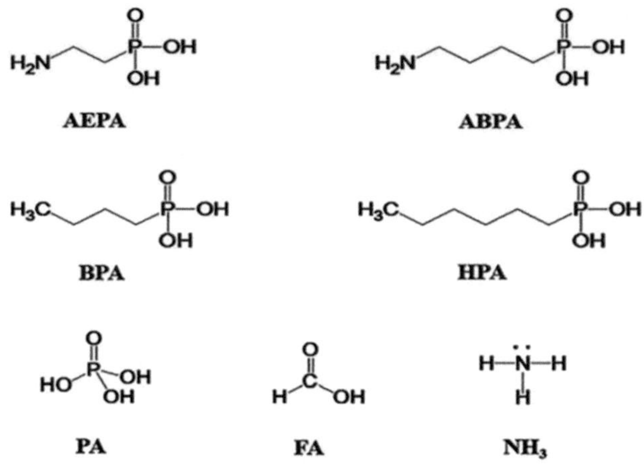


**Figure 5.**
*Chemical structures of the molecules used to form SAMs.*

used alkylphosphic acid by microcontact printing to create patterns on a transparent metal oxide film, wet etched [64]. Robel et al. used thiol-containing organic acids to modify the $TiO_2$ surface to obtain modified Eselenide cadmium (CdSe) quantum dots [65]. There is already a literature on the surface modification of $TiO_2$ by SAM. Furthermore, it has been shown that a stronger bond will form between phosphonate and $TiO_2$ than between carboxylic acids [66, 67]. In this regard, higher DSSC stability should be achieved if phosphonic acid instead of carboxylic acid forms SAM on $TiO_2$ electrodes. Several phosphonic acids with different end groups and molecular lengths were chosen to change the work function of $TiO_2$ to see how the dipole moment affects the DSSC performance. A higher PCE was achieved when $TiO_2$ was modified with amino-containing phosphonic acids [68]. Similar organic acids were selected to adjust the photoanode's surface potential and interfacial properties, enabling electrochemical analysis of SAM effects on DSSC performance. Relevant SAMs include AEPA (2-aminoethylphosphonic acid), ABPA (4-aminobutylphosphonic acid), BPA (1-butylphosphonic acid), HPA (n-hexanylphosphonic acid), PA (phosphoric acid), FA (formic acid), and $NH_3$ (ammonia, 2.0 M ethanol solution), with their structures shown in **Figure 5**.

### *2.2.2 Fixation of the photosensitizer by the self-assembled monolayer*

SAM technology is useful in fixing photosensitizers in DSSCs. Ruthenium (II) polypyridyl complexes represent important dyes in artificial photovoltaic devices because of their ability to convert visible light into chemically or electrically available energy [69]. In particular, complexes of the $[Ru\ (bpy)_3]^{2+}$ family attract great attention in homogenous single-molecule catalysis [70] and, when fixed on a semiconductor metal oxide, also act as photosensitizers in DSSC, for example, titanium dioxide, through the carboxylic acid head group (as shown in **Figure 6**) [71, 72]. Their well-studied chemical stability and biophysical properties are also advantageous for biological applications combined with gold nanoparticles [73, 74], for example, in the field of nanosensing and cellular imaging [75, 76] as well as photothermal therapy [76]. For these applications, the interaction of dyes with gold plays an important role both in their photochemical properties and stability [73].

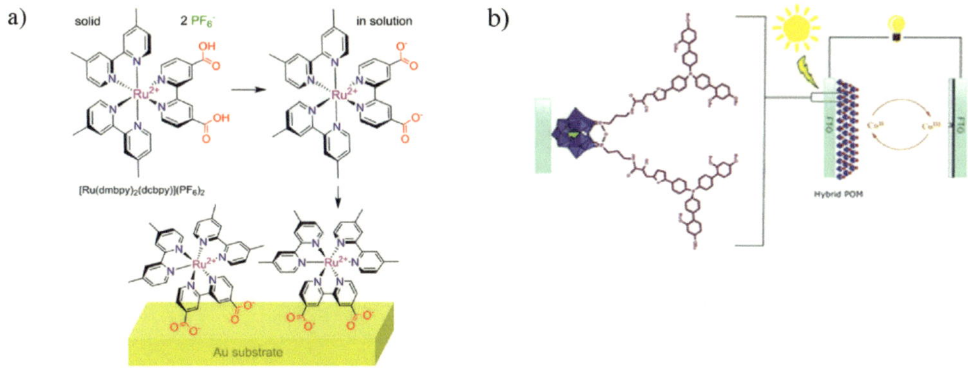


**Figure 6.**
*(a) Schematic representation of the photosensitizer [Ru(dmbpy)$_2$(dcbpy)](PF$_6$)$_2$ (Ru(II) complex) self-assembly on a gold substrate from solution. Self-assembled layers of Ru(II) complex show a loss of PF$_6$ – counterions alongside carboxylate formation. Copyright (2021)* ACS Applied Materials & Interfaces. *(b) schematic representing the covalent bond SAM of D35 dye on the surface of polyoxometalate in DSSCs.*

In another report, self-assembled monolayer formation in DSSCs is achieved through covalent bonding of the D35 organic dye to a Keggin-type hybrid polyoxometalate (hybrid-POM) thin film. The D35 dye is anchored to the Hybrid-POM surface *via* strong amide bonds, ensuring stable attachment and efficient electron transfer. This novel approach enhances dye loading and interfacial charge dynamics, contributing to improved photovoltaic performance in DSSCs [50].

### 2.3 Optimization of the performance of dye-sensitized solar cells by molecular interactions

In DSSCs, the type and mechanism of intermolecular interactions have an important impact on the performance and stability of the cells.

#### *2.3.1 Effect of molecular interactions on the adsorption of the photosensitizer*

DSSC operation relies on a semiconductor anode, a dye photosensitizer, a redox electrolyte, and a conductive cathode. The dye extends the semiconductor's absorption range beyond UV, enabling photon energy conversion into voltage and current [55, 77]. In the art, DSSC is less efficient compared to other mature technologies of other solar cells. However, many researchers emphasize the use of natural species, which are sensitized dyes as light absorbers of DSSC, thus, unaffected grounding procedures, widespread sources that can be obtained at low cost [78].

Enhancing DSSC efficiency relies on photosensitizer dyes absorbing visible light and supplying electrons to $TiO_2$, accelerating electron transfer. Organic, inorganic, and hybrid dyes serve this role. Natural dyes harness solar energy effectively, while synthetic organic dyes offer high efficiency, cost-effectiveness, easy preparation, strong absorption, and environmental sustainability [79, 80]. Inorganic dyes (synthesis) such as the ruthenium groups N3 and N719 have been used as synthetic light collectors in the DSSC and have shown a higher efficiency of up to 13% [1, 2]. Ruthenium-based dyes are costly and scarce, leading to growing interest in natural dyes from fruits, flowers, leaves, seeds, and vegetable skins for DSSCs. These dyes contain pigments that modify color and absorption range, and offer benefits like biodegradability, availability, easy extraction, and enhanced electron transfer. Natural dyes efficiently absorb light and convert it to electricity, while synthetic organic dyes provide high energy conversion, cost-effectiveness, easy preparation, strong absorption, and sustainability [79, 80]. Generally, Flavonoids are natural compounds with phenolic structures and various functional groups like hydroxyl, ethyl, and carboxylic acid, commonly found in plants. They include subclasses such as flavonols, flavonones, isoflavones, and anthocyanins, with examples like quercetin, digitoxin, apigenin, genistein, hesperetin, EGCG, epicatechin, and pine [5]. Colin is the main component of various flowers, plants, and fruits, and has excellent photosensitivity to DSSC performance [4, 6, 8].

### *2.3.2 Effect of molecular interactions on charge transport*

In DSSCs, photons excite electrons in the dye sensitizer, which are then transferred to mesoporous $TiO_2$ nanocrystals and collected by a transparent electrode. The redox mediator reduces the oxidized dye by transferring electrons from the opposite electrode. Today, high-performance DSSC can be obtained using iodide/triiodide I or cobalt $[Co(bpy)_3]^{2+/3+}$ electrolyte acting as a redox mediator. In the past, $I^-/I^{3-}$ electrolytes were the conventional redox means of DSSCs. Moreover, some studies have reported that TBP adsorption onto the $TiO_2$ surface can passivate the $TiO_2$ surface, reducing the interaction of redox substances with $TiO_2$ and thus reducing charge recombination [13, 15]. Other studies have found that the addition of TBP did not cause significant changes in the electron lifetime [14, 16]. While the PDA-enhanced $I^-/I_3^-$ redox mediator is well-studied, few studies explore molecular interactions in the cobalt-PDA-based system. The positively charged cobalt mediator interacts with the negatively charged $I^-/I_3^-$, especially in the presence of TBP with an electron-donating group. Thus, investigating the impact of the cobalt-PDA system on DSC performance is crucial.

## 2.4 Effects on stability and longevity

SAMs and molecular interactions significantly affect the stability and lifespan of DSSCs. Suitable SAMs optimize the electrode's work function, improving charge transfer and enhancing photogenerated carrier separation. This boosts photoelectric conversion efficiency and stability. For example, the Bz-Ph_p_PACz SAM bilayer structure enhances hole extraction efficiency, improving DSSC performance and stability. Additionally, SAMs provide a protective barrier against environmental factors like humidity and oxygen, with hydrophobic or antioxidant SAMs extending battery lifespan. In DSSCs, π-π stacking interactions between dye molecules and SAMs on the electrode surface improve light absorption, charge transport, and stability under light. Weaker interactions like hydrogen bonds and van der Waals forces also play a crucial role in maintaining cell integrity, enhancing dye adsorption, and reducing dye degradation, thus extending the cell's lifespan. Additionally, intermolecular charge transfer interactions promote faster charge separation and transmission, improving open-circuit voltage and short-circuit current. However, factors like the cell preparation process, material purity, and stability must also be considered for optimal performance and longevity.

# 3. Applications in organic photovoltaics (OPVs)

The significant application of SAMs in Organic Photovoltaics (OPVs) in solving the efficiency and stability challenges has attracted the researchers attention to further study.

### 3.1 Organic photovoltaics (OPVs): An overview

OPVs are devices that convert light energy into electricity using organic materials like polymers as conductors. They are seen as a promising alternative to traditional solar cells due to their abundance, cost-effectiveness, and environmental benefits. As third-generation solar cells, OPVs utilize multiple bandgaps, with electrodes collecting charge carriers generated by sunlight [81].

### 3.2 Importance of nanoscale engineering in OPVs

One key nanoscale process in organic solar cells is self-assembly, which enhances light absorption, promotes effective charge separation, and minimizes energy losses. Self-assembly techniques are employed to create active layers with diverse morphologies and to form protective layers that shield the cells from environmental factors. With ongoing advancements in this area, there is a strong expectation for the development of organic solar cells that combine high efficiency with low production costs [82].

### 3.3 SAMs in OPVs

SAMs are used to modify the physical and chemical properties of substrate surfaces, significantly improving the electrical performance and stability of OPVs. By optimizing charge transfer and band alignment, SAMs enhance interfacial adhesion between organic layers and electrodes, facilitating better charge extraction and transport. SAMs also help mitigate both extrinsic degradation (e.g., moisture and oxygen exposure) and intrinsic degradation (e.g., internal layer changes), acting as a protective barrier. This improves device efficiency and extends operational lifespan, making SAMs crucial for advancing OPV commercialization [83].

### 3.4 Phase separation in OPVs

In order to improve the efficiency of SAMs-based OPVs, controlling the phase separation and understanding the techniques to do so is necessary.

#### *3.4.1 Phase separation*

Numerous studies have enhanced OPV performance by optimizing phase separation in the active layer. In solution-processed OPVs, spin coating is commonly used to fabricate the active layer. Factors like material properties, the donor-acceptor ratio, and pre- and post-treatment processes can significantly impact phase separation. Although current research shows promising results, further optimization of phase separation is needed for better performance in OPVs [84].

#### *3.4.2 Phase separation controlling: Techniques*

In order to improve phase separation, different strategies have been studied in SAM-based OPVs, including:

##### *3.4.2.1 Layer-by-layer (LbL) processing*

The LbL method enables sequential deposition of donor and acceptor materials, allowing precise control over the active layer's morphology and interface. By promoting vertical phase separation, LbL enhances exciton diffusion, reduces charge recombination, and improves charge transport. This results in higher device efficiency, making LbL an effective approach for developing high-performance OPVs, especially in applications with thicker active layers [85].

#### *3.4.2.2 Solvent additives*

Utilizing solvent additives like DIO (1,8-diiodooctane) in SAMs-based OPVs during the film formation significantly affects the vertical phase separation by creating a strong vertical composition gradient [86].

#### *3.4.2.3 In situ photo-polymerization*

*In situ* polymerization during film formation enables controlled vertical phase separation between donor and acceptor materials. This technique improves the morphology of bulk heterojunction solar cells, optimizing exciton dissociation and charge transport. By achieving well-defined vertical separation, it enhances power conversion efficiency by up to 20% and addresses challenges related to active layer thickness [87].

## 3.5 Light management strategies

Light absorption enhancement is an effective step toward improving the SAMs-based OPVs. These devices can be improved by utilizing nanostructures and employing plasmonic nanostructures in order to increase light scattering and trapping, and thus the light absorption.

### *3.5.1 Structured nanostructures for enhanced light absorption*

Self-assembly methods enable the creation of structured nanostructures, such as silica nanoparticle arrays and core-shell nanospheres, on polymer substrates. These nanostructures effectively scatter light, increasing the optical path length in OPVs' active layers. For instance, self-assembled silica nanoparticle layers on flexible substrates enhance light absorption by increasing transmission haze while maintaining acceptable transmission losses [88].

OPVs use a bulk heterojunction structure, where the electron donor and acceptor are mixed to capture light. While increasing the thickness of this layer can enhance light absorption, it also makes charge carrier extraction more difficult. As a result, the photoactive layer is typically kept thin—less than 100 nm—which limits sunlight capture and reduces PCE. Light trapping techniques, like plasmonic nanoparticles and photonic crystals, help improve absorption but may introduce optical losses and defects. Self-assembly methods offer a promising solution, enabling well-organized nanostructures that enhance light trapping while maintaining charge extraction efficiency, thus improving OPV performance [89].

### *3.5.2 Plasmonic nanostructures for enhanced light absorption*

Self-assembly methods for plasmonic monolayers focus on reducing parasitic absorption and optimizing nanoparticle spacing. By carefully arranging nanoparticles, exciton quenching is minimized while light scattering is enhanced within the active layer. This approach significantly improves the optical properties of OPVs, advancing solar energy conversion. A study with silver nanoparticle arrays, specifically nanoparabolas, showed notable improvements in light absorption at visible and near-infrared wavelengths. These plasmonic structures enhance the electric field at the metal-dielectric interface, leading to a 26.41% increase in JSC and a 26.37% boost in PCE, outperforming planar OPVs [89].

### 3.6 Charge transport enhancement

There are two important factors in enhancing the charge transport in SAMs-based OPVs including exciton diffusion length, interface engineering, morphology, nanowire networks management, and energy-level alignment.

#### *3.6.1 Exciton diffusion length considerations*

One of the critical factors influencing the performance of OPV cells, particularly in the context of self-assembly techniques, is exciton diffusion (**Figure 7**). Excitons, generated by light absorption, are subsequently separated into free charge carriers when they transfer to the interface between donor and acceptor materials. The emission length of excitons plays a pivotal role in determining the efficiency of this process. In conventional organic semiconductors, exciton diffusion lengths typically range from 5 to 20 nm [90, 91]. Self-assembly techniques offer a solution to the constrained emission length by creating ordered structures with minimal defects in the photoactive layer. This optimization of molecular arrangement enhances exciton diffusion, improving exciton emission and overall performance. By addressing the limitations of short propagation lengths, self-assembly aligns with strategies to maximize exciton lifetimes and reduce charge generation losses, leading to more efficient OPVs [92].

#### *3.6.2 Interface engineering for efficient charge transfer*

SAMs are crucial for interface engineering in OPVs, enhancing charge transfer by modifying adjacent surfaces. These densely packed, spontaneously formed 2D films consist of anchoring groups, linkers, and terminal groups that interact with the environment. By incorporating dipolar groups, SAMs can modify the work function of the substrate, allowing precise control over energy-level alignment and reducing injection barriers. This improves charge transport and boosts the overall performance of OPVs [93].

#### *3.6.3 Controlled morphology and nanowire networks for improve charge transfer*

Morphology is key to OPV performance, influencing charge generation, transfer, and collection. Self-assembly techniques, particularly supramolecular interactions like hydrogen bonding and π-π stacking, optimize morphology by stabilizing nanostructures and enhancing charge transfer pathways [94, 95]. Another technique involves the use of self-assembly to form nanowires. Nanowires, due to their high surface-to-volume ratio, are effective for charge collection and separation [95, 96]. Techniques such as solvent evaporation or gradual cooling of saturated solutions can induce self-assembly in nanowires. The choice of processing solvents and their interaction with the side chains of donor and acceptor materials significantly impacts morphology [97].

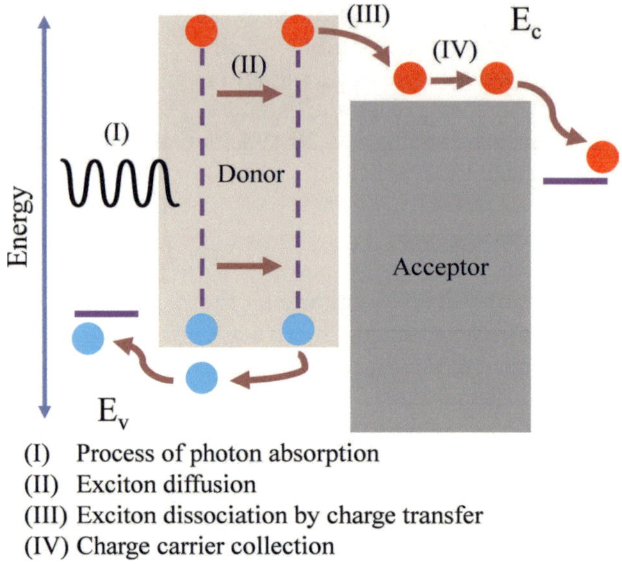


**Figure 7.**
*Charge transfer mechanisms in photovoltaic systems.*

### *3.6.4 Energy-level alignment*

When materials with different electronic energy levels meet, an energy barrier can hinder charge carrier movement. SAMs help modulate energy-level alignment by adjusting the work function (WF) of metals and semiconductor surface potentials. SAMs with electron-withdrawing groups (e.g., -CF3) increase WF, while electron-donating groups (e.g., tert-butyl) decrease it. This tuning reduces energy barriers for charge transport and improves device performance by enhancing ohmic contact at electrodes, increasing $V_{OC}$, $J_{SC}$, and FF in solar cells [98–100].

## 4. Applications in quantum dot photovoltaics (QDPVs)

### 4.1 Introduction to QDPV

Quantum dots (QDs) are semiconductor nanocrystals with tunable electronic and optical properties due to quantum confinement effects. In QDPV devices, QDs absorb photons, generating electron-hole pairs that are separated and transported to electrodes, producing an electric current. A typical QDPV structure includes a transparent conductive electrode, a QD active layer for light absorption, charge transport layers (CTLs) for efficient carrier extraction, and a back electrode, optimizing photovoltaic performance.

Over the past two decades, semiconductor QDs have garnered significant attention due to their unique photovoltaic advantages stemming from their zero-dimensional nature. Self-assembled quantum dots (SAQDs) offer enhanced carrier generation through tailored QD energy levels or band structures [101, 102]. Expanding the spectral range for light excitation can enhance thermodynamic conversion efficiency. However, practical photovoltaic performance depends on carrier transport and collection mechanisms. A thorough understanding of these processes is essential for designing high-PCE solar cell architectures. Advancements in SAQD growth techniques are crucial for high-quality solar cells with controlled heterointerfaces and effective surface passivation. Recent progress in SAQD-based solar cells highlights their potential, positioning QDPV as a promising technology for sustainable energy development.

#### *4.1.1 Overview of self-assembly technology and its importance in QDPVs*

Self-assembly technology utilizes weak intermolecular forces to spontaneously form highly ordered nanostructures, enabling precise molecular or nanoparticle arrangement without significant external intervention. In QDPV devices, self-assembly technology ensures a uniform quantum dot active layer by minimizing aggregation and optimizing inter-particle spacing. This precise arrangement enhances photon absorption and charge generation, significantly improving device efficiency. Ordered quantum dot films formed *via* self-assembly demonstrate substantially higher light absorption compared to disordered films. The self-assembly of quantum dots (QDs) occurs due to the relaxation of strain energy in epitaxial systems, where the lattice parameters of the deposited material significantly exceed those of the underlying substrate [103]. This process enhances the stability and mobility of QDs while suppressing thermal flow and nonradiative transitions. Moreover, post-self-assembly tuning of QD size expands their application potential in photovoltaics, diodes, and biomedical fields [104–107]. Additionally, various types of QDs and their assemblies offer significant practicality for studies on photonic crystal coupled emission (PCCE) [107]. Recent research has observed self-assembly phenomena in diverse QD materials. In 2009, García-Ruiz et al. reported that barium carbonate or strontium carbonate precipitation in silica-rich alkaline environments forms crystalline aggregates, which are classified as pure inorganic self-assembled nanocrystals [108]. In 2019, Grünwald et al. utilized molecular dynamics simulations to investigate the self-assembly of these nanocrystals under varying ligand lengths and solvent conditions, as illustrated in **Figure 8a** [109]. Their findings revealed that subtle differences in nanoparticle shape, ligand length and coverage, and solvent conditions can lead to markedly different self-assembled superstructures due to minor variations in ligand interaction free energy. In 2024, Vaynzof et al. [110] demonstrated, for the first time, lead sulfide (PbS) quantum dot solar cells employing carbazole- and dibenzothiophene-based self-assembled molecules as hole transport layers. Among the studied molecules, Br-2EPT exhibited superior photovoltaic performance, achieving a power conversion efficiency of up to 6.3%. These results underscore the significant potential of self-assembled molecules as hole transport layers in inverted p-i-n PbS QD solar cells.

### 4.2 Self-assembly for creating uniform quantum dot thin films

Self-assembly technology is a key method for fabricating uniform QD thin films in QDPV applications. Chemical bonding-based self-assembly utilizes reactions between QD surface functional groups and substrates or ligands, enabling precise film formation. For instance, amidation between carboxylated QDs and amino-functionalized substrates facilitates uniform single or multilayer film growth. This approach ensures controlled QD distribution, enhancing film quality and device performance.

#### *4.2.1 Molecular beam epitaxy (MBE)*

MBE is a key technique for fabricating high-quality semiconductor QDs, thin films, and heterojunctions. This process involves the precise deposition of atoms onto a solid

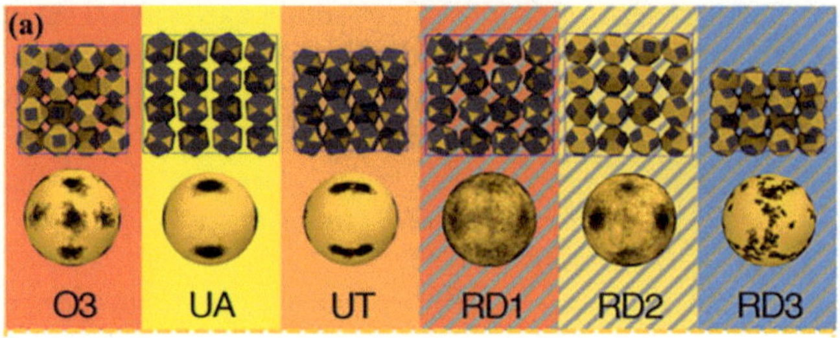


**Figure 8.**
*(a) Images of various quantum dots after molecular dynamics simulation [109].*

substrate, where they adsorb, migrate, and nucleate to form uniform material layers, enabling advanced semiconductor device development [111]. In 2020, Sautter et al. conducted an in-depth investigation into the impact of strain on the self-assembly of QD [112]. Their study examined the impact of compressive and tensile strain on self-assembled QDs, offering key insights for designing QDs with tailored properties. In 2023, Vallejo et al. used MBE to successfully grow $In_{0.5}Ga_{0.5}As$ QDs on a GaSb(111) A surface, achieving high structural integrity and optical quality [113]. With a 4% tensile strain, $In_{0.5}Ga_{0.5}As$ deposited on GaSb(111)A enabled QD self-assembly. The QDs remained fully strained and defect-free below a threshold size, while residual tensile strain allowed tunable mid-infrared emission in InGaAs(Sb)/GaSb(111)A QDs.

### *4.2.2 Vertical sedimentation method*

The vertical sedimentation method was first introduced by Nagayama in 1995 [114]. This technique leverages the interaction between capillary forces and surface tension to achieve the orderly assembly of colloidal particles. They found that particle self-assembly occurs abruptly as solvent evaporation thins the liquid layer, forming a hexagonal lattice structure. This effect intensifies with higher solvent concentrations. They suggested that controlling the liquid layer's thickness can induce self-assembly, with equilibrium thickness determined by osmotic and capillary pressures. This method yields high-quality samples with fewer defects and more uniform thickness.

In 2018, Mukai et al. enhanced the performance of solar cells by utilizing PbS QD superlattices prepared *via* the deposition method [115]. By controlling vapor pressure, Mukai et al. regulated solvent evaporation and sedimentation time of PbS quantum dots, leading to the formation of larger superlattices. A solar cell using this QD superlattice film as the absorption layer showed improved short-circuit current density and photoelectric conversion efficiency. Reducing the deposition rate nearly doubled the short-circuit current density, highlighting the impact of sedimentation time on interface uniformity. Additionally, they used a micro-hole array template for PbS QD deposition, forming a superlattice film [115]. As shown in **Figure 9**, the high-order emission lifetime on the templated surface was more than twice that of a planar substrate. This extended lifetime indicates that the template increases the grain size of individual QD superlattices, enhancing their structural integrity and performance.

### *4.2.3 Chemical vapor deposition (CVD)*

Chemical vapor deposition (CVD) is a key technique for synthesizing nanomaterials, known for its low cost, simplicity, high throughput, and scalability. In CVD,

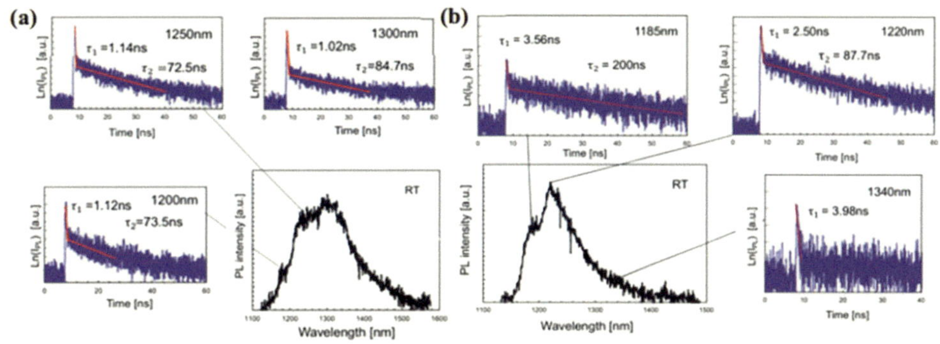


**Figure 9.**
*(a) μ-PL spectra and decay curves of QD films prepared for 20 minutes on a flat substrate. The vertical axis of the decay curve represents the natural logarithm of the photoluminescence (PL) intensity. (b) μ-PL spectra and decay curves of QD films prepared over 7 days on a templated surface. The vertical axis of the decay curve also represents the natural logarithm of the PL intensity [115].*

gaseous precursor molecules decompose at elevated temperatures, and the resulting reactive species react with the substrate to form solid thin films. This method offers precise control over film composition, thickness, and morphology, making it versatile for various applications.

In 2016, Yan et al. first synthesized fluorescent carbon quantum dots (C-CQDs) with a graphite-like structure using the chemical vapor deposition (CVD) method [115]. The synthesized C-CQDs exhibited excellent crystalline graphite properties. Polymer solar cells were fabricated using C-CQDs as the electron transport layer (ETL), enhancing electron injection and reducing interfacial resistance. This resulted in superior device performance compared to those using lithium fluoride as the ETL. Additionally, the devices with C-CQDs showed improved thermal stability. The J-V curve and external quantum efficiency (EQE) spectrum of the P3HT:PC61BM devices confirmed these enhancements (**Figure 10**).

### *4.2.4 Ligand exchange method*

Quantum dots (QDs) are typically synthesized through solution-phase methods, allowing precise control over their size, shape, and composition. Ligands passivate the particle surface during nanoparticle formation, but long-chain organic ligands can limit inter-particle interactions. Ligand exchange can shorten these chains, enhancing QD self-assembly and improving performance. This process, which uses short-chain organic molecules or inorganic ions, optimizes QD ordering and is applied before or after thin-film deposition. However, high electronic resistivity and ligand hydrophobicity remain challenges for device performance. Increasing QD size or incorporating smaller QDs can create more ordered and diverse mixed lattices, enhancing functionality.

In 2020, researchers demonstrated a universal and scalable method to enhance the light absorption and exciton coupling of QDs through self-assembly [116]. QDs were self-assembled into colloidal superlattices using emulsion templates *via* ligand exchange. These superlattices exhibited extended resonant optical behavior, with absorption efficiency in the visible light range increasing by more than two orders of magnitude compared to dispersed QDs. In 2021, Zheng et al. introduced a passivation strategy for QD self-assembly to improve QD coupling [117]. This method utilized ammonium iodide for ligand exchange, effectively replacing oleic acid (OA) ligands

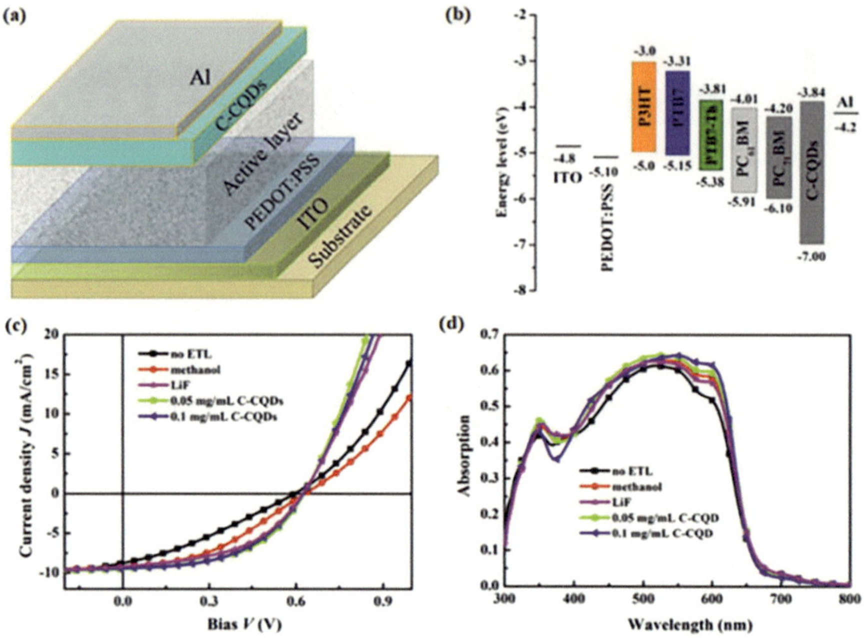

**Figure 10.**
*(a) Structure of the polymer solar cell. (b) Energy-level diagram of each functional layer in the PSC. (c) J–V curve of the P3HT:PC61BM device. (d) External quantum efficiency (EQE) spectrum of the P3HT:PC61BM device [115].*

with iodide ions ($I^-$). The resulting charge balance on the QD surface ensured high stability of the QD ink. Subsequently, a $PbI_2$ passivation layer was assembled on the QD surface, forming QD-$PbI_2$ structures. The self-assembled $PbI_2$ layer not only passivated the QD surface but also acted as a "charge bridge" between adjacent QDs, enhancing charge transport. As a result, the power conversion efficiency of solar cells based on these self-assembled QDs increased by 12.3%. This work provided a simple and effective approach to improving the electronic coupling of QDs. In 2022, Coropceanu et al. reversibly self-assembled colloidal nanocrystals of gold, platinum, nickel, lead sulfide, and lead selenide with conductive inorganic ligands into superlattices [118]. These superlattices exhibited optical and electronic properties consistent with strong electronic coupling between the constituent nanocrystals. This achievement overcame the limitations imposed by insulating organic surface ligands on the collective electronic states in ordered nanocrystal assemblies.

### 4.3 Charge transport mechanism

In a QDPV device, quantum dots absorb photons of specific wavelengths due to their quantum size effect. The photon energy must match the QD's bandgap for electrons to be excited, creating photogenerated carriers. By controlling QD size, composition, and surface ligands, their bandgaps can be tailored to absorb a broader range of the solar spectrum. Electron transport occurs *via* hopping between adjacent QDs, and this efficiency is influenced by factors like inter-dot distance and electron cloud overlap. Self-assembly techniques can create a uniform, densely packed QD film, reducing distances and increasing overlap, which enhances electron transport efficiency.

#### *4.3.1 Optimization of charge transport pathways via self-assembly structures*

Self-assembly technology enables the formation of well-ordered QDs, reducing inter-dot distances and enhancing electron cloud overlap. This arrangement facilitates easier electron hopping through quantum tunneling, significantly improving electron transport efficiency. In early PbS colloidal CQD solar cells, simple Schottky structures were used to study the basic properties of CQD materials and the effects of surface treatments on photovoltaic performance. These devices typically consisted of a CQD active layer between a transparent electrode and a metal electrode, with some interface modifications [119]. However, due to the poor hole injection barrier at the ITO/PbS CQD interface and Fermi level pinning caused by defect states at the metal/PbS CQD interface, the open-circuit voltage ($V_{OC}$) was relatively low, limiting device performance [120]. To date, the highest PCE achieved using this structure is only 5.2% [121]. To address the limitations of Schottky junction CQD solar cells, the most widely adopted passivation technique involves introducing SAMs or ultrathin organic semiconductors at the interface between the electron transport layer and the CQD layer. For instance, as shown in **Figure 11**, Kim et al. utilized benzoic acid-functionalized SAMs on ZnO ETLs, adjusting the energy alignment at the rectifying interface by modifying the dipole moment of the SAM structure [105]. Additionally, SAM anchoring exhibits high stability against subsequent chemical treatments required for CQD solid fabrication, achieving a PCE of 10.7% at that time. The same 4-aminobenzoic acid SAM was later employed at the interface of ordered bulk ZnO nanowire/PbS CQD heterojunctions, significantly enhancing the voltage and fill factor of the photovoltaic devices [122].

### 4.4 Light collection principle

The separated electrons and holes need to be transported within the QD system to the electrodes to form a current. The transport of electrons and holes between QD can occur through tunneling effects or hopping mechanisms. Factors such as the spacing between QD, surface ligands, and the arrangement of quantum dots all influence the efficiency of carrier transport. For instance, reducing the thickness of ligands between QD or using ligands with high conductivity can shorten the carrier transport path and enhance the transport efficiency.

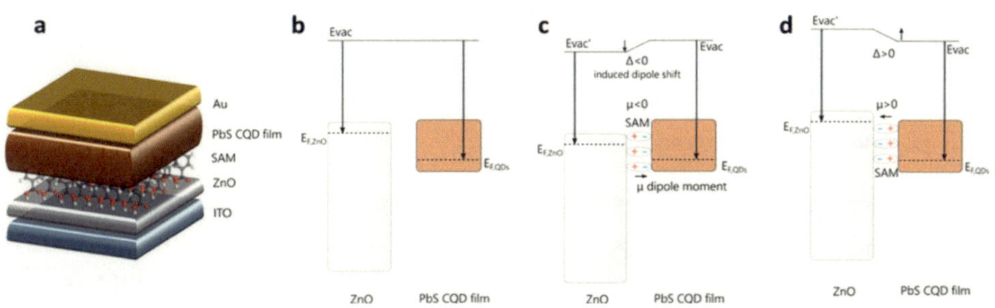


**Figure 11.**
*Band alignment in CQD photovoltaics. (a) Schematic illustration of the prototype colloidal quantum dot (CQD) solar cell structure, featuring a self-assembled monolayer (SAM) between the ZnO and PbS CQD films. (b) Energy-level diagrams of the original ZnO-PbS junction and the SAM-modified interface with (c) positive or (d) negative dipole moment molecules. The induced shift in the vacuum level (Δ) is determined by the density, dipole moment, and orientation of the SAM constituent molecules [105].*

#### *4.4.1 Light scattering in self-assembled quantum dot films*

Semichaevsky and Johnson [123] utilized a multiscale carrier transport model to simulate p-i-n solar cells incorporating InAs/GaAs quantum dots (QDs). Their results demonstrated that while high planar density stacks of InAs QD arrays in the solar cells can generate photocurrent by absorbing photons with energy below the bandgap of bulk GaAs, they also inhibit the transport of photogenerated carriers resulting from the absorption of higher-energy photons. The confinement potential induces quantum scattering, leading to longer carrier travel paths and increased nonradiative (NR) recombination within the intrinsic region of the cell [124]. Additionally, reflected carriers form additional space charges, which reduce the built-in field in the heterostructure region.

## 5. Emerging trends in tandem photovoltaic architectures

In recent years, significant progress has been made in solar cell technology, with tandem solar cells becoming a research hotspot due to their breakthrough efficiency improvements. Tandem solar cell (TSC), or multi-junction solar cell, is a kind of special PV device, in which two or more subcells with complementary absorption are connected and stacked in series or parallel. This chapter will delve into the role of self-assembly strategies in three types of perovskite tandem solar cells, analyzing their unique advantages in enhancing photovoltaic efficiency and their future application prospects.

### 5.1 The overview of self-assembly monolayer in tandem photovoltaics

In a series multi-junction solar cell, multiple layers of materials with different bandgaps are stacked on top of each other. With a high degree of molecular integration and interface control, SAMs can be used to improve the interface quality between these material layers, promote charge transfer, and reduce interface energy loss, playing a crucial role in improving the efficiency, stability, and lifetime of series photovoltaic devices. In recent years, research based on hole-selective SAMs has boosted (**Figure 12a, b**), which have achieved PCEs of 26.81 and 32.8% for single-junction and monolithic tandem PSCs, respectively. The outstanding PCE records are among the high-level perovskite photovoltaics, indicating the superiority of SAMs. The chemical structures of representative SAMs are shown in **Figure 12c** [12].

The application of SAMs in perovskite tandem devices and the corresponding PCEs are summarized in **Table 1** [125].

### 5.2 All-perovskite tandem solar cells

All-perovskite TSCs consist of a lead-based WBG (1.75–1.8 eV) perovskite top subcell and a tin lead (Sn-Pb) LBG (1.2–1.3 eV) perovskite bottom subcell, consisting of two terminal (monolithic or 2-T, **Figure 13a**) and four-terminal (4-T, **Figure 13b**) architecture. Compared with single-junction perovskite solar cells (PSCs), it has a higher power conversion efficiency (PCE). In addition, all-perovskite series solar cells have the advantages of low temperature solution preparation and low manufacturing cost, and have received widespread attention, and are considered to be one of the most potential next-generation high-performance thin-film photovoltaic technologies.

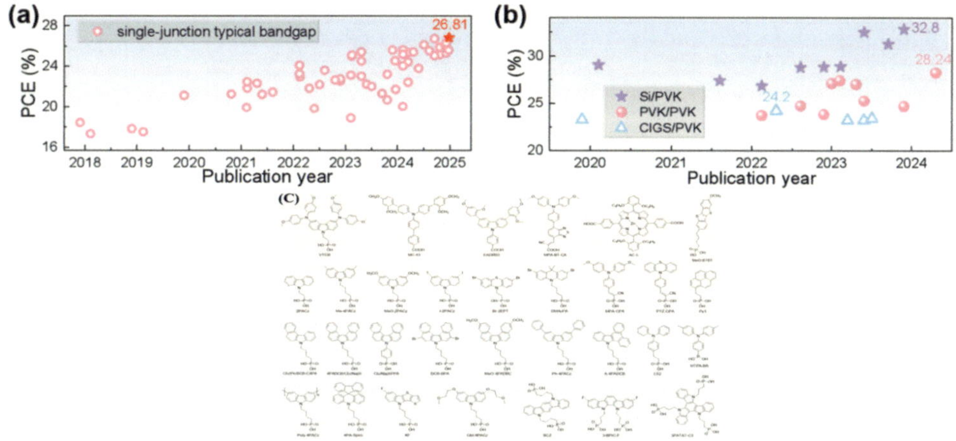


**Figure 12.**
*Progress of hole-selective SAMs in PSCs. (a, b) PCE records of PSCs based on SAMs, including single-junction inverted PSCs with a typical bandgap (1.5–1.65 eV) perovskite and monolithic tandem devices; (c) The chemical structures of representative SAMs. Reproduced with permission [12]. Copyright 2025,* Chemical Engineering Journal.

The structure of SAMs usually consists of three segments, including the functional group, spacer group, and anchoring group, of which the functions are transporting holes and passivating defects, linking functional and anchor group, and attaching to the metal oxide substrate, respectively.

High-bandgap subcells have been identified as the primary cause of $V_{OC}$ losses in numerous tandem devices, as reported in the literature. To address this issue, Thiesbrummel et al. implemented a triple optimization strategy for high-bandgap perovskites [126]. The strategy involved substituting HTL PTAA with 2PACz, incorporating oilamine into the perovskite layer, and introducing a thin LiF layer between the perovskite and ETL to enhance the performance of the high-bandgap subcell (**Figure 14A**). The optimized high-bandgap perovskites were used to fabricate efficient all-perovskite tandem solar cells. The steady-state efficiencies obtained for the 1.80 eV/1.27 eV, 1.85 eV/1.27 eV, and 1.88 eV/1.27 eV bandgap combinations were 23.4, 23.7, and 21.5%, respectively (**Figure 14B**). Lai et al. proposed a comprehensive optimization strategy for various components of the device. They employed 2PACz as the HTL for 1.77 eV triple cation perovskite devices, effectively suppressing $V_{OC}$ losses at the HTL/perovskite interface [127]. The researchers achieved a $V_{OC}$ of 1.29 V and developed a near-infrared transparent wide-bandgap (WBG) PSC with a PCE of 15.1% on flexible substrates. Notably, the high $V_{OC}$ corresponded to a record-low $V_{OC}$ deficit (480 mV) for the perovskite, considering a bandgap of approximately 1.80 eV. Moreover, by combining flexible narrow-bandgap (NBG) (1.24 eV) PSCs, they successfully fabricated flexible all-perovskite four-terminal (4 T) tandem solar cells with impressive PCEs of 22.6 and 23.8% for different configurations (**Figure 14C**).

Wang et al. reported a universal SAM-based HTL that rivaled the high-performance PTAA and PEDOT: PSS in WBG PSCs. The SAM, based on the structure of (4-(10-bromo-7H-benzo[c]carbazole-7-yl)butyl)phosphonic acid (BCBBr-C4PA), combined an asymmetric conjugated backbone and bromination strategy in molecular design, thereby enhancing solubility and increasing the dipole moment (**Figure 14D**) [128]. The BCBBr-C4PA HTL exhibited negligible light absorption and a lower HOMO energy level, promoting efficient interfacial charge transfer and

| SAMs | Configuration | $V_{OC}$ (V) | $J_{SC}$ (mA cm$^{-2}$) | FF (%) | PCE (%) |
|---|---|---|---|---|---|
| MeO-2PACz | CIGSe/SAM/perovskite(triple cation)/$C_{60}$/BCP/Cu | 1.68 | 19.17 | 71.9 | 23.26 |
| MeO-2PACz | CIGSe/CdS/intrinsic zinc oxide (IZO)/aluminum-doped zinc oxide (AZO)/$NiO_x$:Cu + SAM/perovskite/$C_{60}$/$SnO_2$/IZO/Ag | 1.70 | 19.47 | 71.0 | 23.5 |
| MeO-2PACz | ITO/perovskite subcell (1.85 eV)/$SnO_x$/organic subcell (1.33 eV)/Ag | 2.15 | 14.0 | 80.0 | 24.0 |
| 2PACz | ITO/2PACz/WBG perovskite (1.85 eV)/$C_{60}$/AZO/$SnO_x$$InO_x$/PEDOT:PSS/NBG perovskite/(1.27 eV)/$C_{60}$/BCP/Cu | 2.0 | 14.8 | 80.0 | 23.7 |
| 2PACz | ITO/2PACz/WBG perovskite (1.77 eV)/PCBM/ZnO/IZO/NBG PSCs | 2.1 | 15.1 | 75.1 | 23.8 |
| 2PACz and MeO-2PACz | PET/ITO/MB-NiO/WBG perovskite (1.75 eV)/$C_{60}$/ALD-$SnO_2$/Au/PEDOT:PSS/NBG perovskite (1.22 eV)/$C_{60}$/BCP/Cu | 2.0 | 15.8 | 78.3 | 24.7 |
| BCBBr-C4PA | PET/ITO/BCBBr-C4PA/WBG perovskite (1.77 eV)/$C_{60}$/ALD-$SnO_2$/Au/PEDOT:PSS/NBG perovskite/$C_{60}$/BCP/Cu | 1.26 | 24.35 | 85.5 | 26.24 |
| 4PADCB | ITO/4PADC/WBG perovskite (1.77 eV)/$C_{60}$/$SnO_2$/IZO/PEDOT:PSS/NBG perovskite (1.25 eV)/$C_{60}$/$SnO_2$/Cu | 2.12 | 15.42 | 82.6 | 27.01 |
| MeO-2PACz and Me-4PACz | ITO/SAM/1.75 eV perovskite/LiF/$C_{60}$/33 nm $SnO_x$/1 nm Au/PEDOT:PSS/1.25 eV perovskite/$C_{60}$/BCP/Ag | 2.20 | 15.3 | 80.8 | 27.1 |
| MeO-4PACz | ITO/$NiO_x$/MeO-4PACz/WBG perovskite (1.79 eV)/$C_{60}$/ALD-$SnO_2$/Au/PEDOT:PSS/NBG perovskite (1.22 eV)/$C_{60}$/ALD-$SnO_2$/Ag | 2.19 | 15.1 | 83.1 | 27.4 |
| 2PACz and MeO-2PACz | Si/ITO/$NiO_x$/mixture/perovskite (1.66 eV)/$CF_3$PhAI/$C_{60}$/BCP/Ag | — | — | — | 26.59 |
| 2PACz | Si/ITO/2PACz/perovskite (1.68 eV)/$C_{60}$/$SnO_2$/IZO/Ag | 1.84 | 19.6 | 76.0 | 27.4 |
| 2PACz | Si/ITO/$NiO_x$/2PACz/perovskite/$C_{60}$/BCP/Ag | — | — | — | 27.6 |
| MeO-2PACz | Si/ITO/MeO-2PACz/perovskite (1.63 eV)/$C_{60}$/BCP/$SnO_2$/Ag | 1.80 | 19.3 | 81.9 | 28.5 |
| 2PACz and MeO-2PACz | Si/ITO/mixture/$Cs_{0.05}(FA_{0.74}MA_{0.26})_{0.95}Pb(I_{0.74}Br_{0.26})_3$/LiF/$C_{60}$/BCP/ITO/Ag | 1.86 | 19.4 | 79.6 | 28.8 |
| Ph-2PACz | Si/ITO/Ph-2PACz/$Cs_{0.15}FA_{0.65}MA_{0.2}Pb(I_{0.8}Br_{0.2})_3$ (1.67 eV)/LiF/$C_{60}$/BCP/Cu | 1.91 | 19.1 | 79.1 | 28.9 |

DOI: http://dx.doi.org/10.5772/intechopen.1012292

| SAMs | Configuration | $V_{OC}$ (V) | $J_{SC}$ (mA cm$^{-2}$) | FF (%) | PCE (%) |
|---|---|---|---|---|---|
| Me-4PACz | Si/ITO/MeO-2PACz/perovskite (1.68 eV)/LiF/$C_{60}$/$SnO_2$/Ag | 1.90 | 19.26 | 79.52 | 29.15 |
| Me-2PACz | Si/$MgF_2$/glass/ITO/MeO-2PACz/perovskite (1.65 eV)/$C_{60}$/$SnO_2$/ITO/$MgF_2$ | — | — | — | 29.8 |
| Me-4PACz | Si/ITO/Me-4PACz/$Cs_{0.18}FA_{0.82}Pb(I, Br)_3$/$C_{60}$/$SnO_x$/IZO/Ag | 1.91 | 20.47 | 79.92 | 31.25 |
| Me-4PACz | Si/TCO/Me-4PACz/$Cs_{0.05}(FA_{0.77}MA_{0.23})_{0.95}Pb(I_{0.77}Br_{0.23})_3$ (1.68 eV)/LiF/$SnO_2$/IZO/LiF/Ag | 1.98 | 20.52 | 79.99 | 32.5 |

**Table 1.**
*Application of SAMs in perovskite tandem solar cells and the corresponding PCE summary. Reproduced with permission [125]. Copyright 2024,* Interdisciplinary Materials.

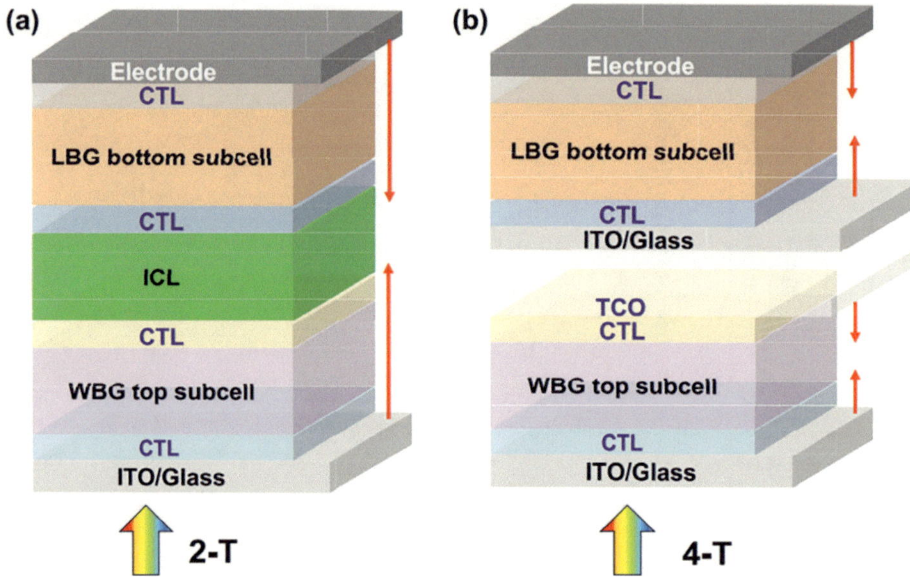


**Figure 13.**
*Schematic device structure of an all-perovskite tandem solar cell. (a) 2-T tandems and (b) 4-T tandems. CTL, charge transport layer; ICL, interconnecting layer.*

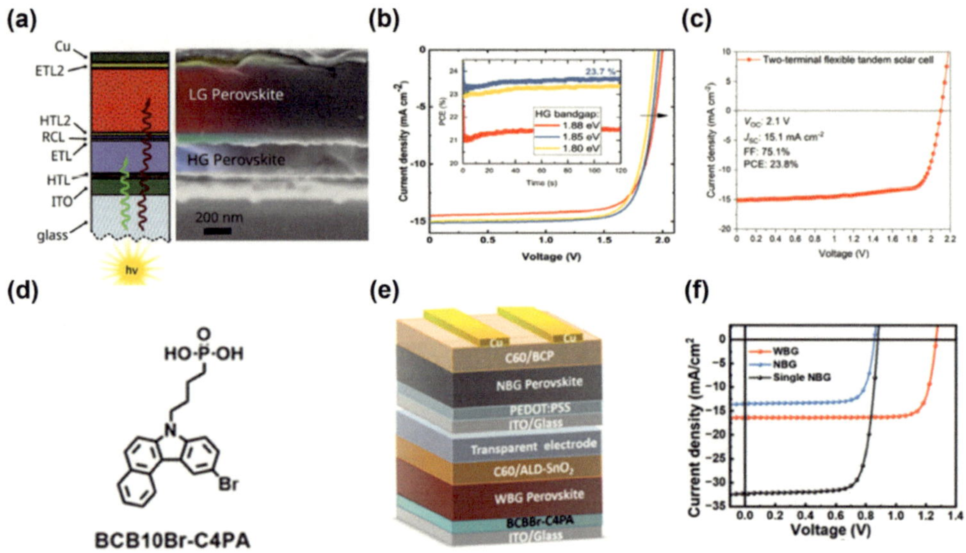


**Figure 14.**
*(A) Schematic diagram of the all-perovskite tandem structure and corresponding cross-sectional images. Reproduced with permission. (B)* J–V *curves for tandem devices with three different perovskite bandgaps. Reproduced with permission. Reproduced with permission [126]. Copyright 2023,* Advanced Energy Materials. *(C)* J–V *curves for two-terminal all-perovskite flexible tandem solar cells. Reproduced with permission [127]. Copyright 2022,* Advanced Energy Materials. *(D) Molecular structure of BCBBr-C4PA, (E) schematic structure, and (F) J-V curves of a 4 T all-perovskite tandem device in the presence of BCBBr-C4PA. Reproduced with permission [128]. Copyright 2023,* Advanced Energy Materials.

suppressing nonradiative recombination losses. The WBG PSCs based on the BCBBr-C4PA SAM achieved a maximum PCE of 18.63%, with the efficiency remaining above 90% after 250 h of continuous operation. By stacking the optimized WBG PSC with an NBG PSC as the bottom cell, a 4 T all-perovskite tandem solar cell achieved a remarkable PCE of 26.24% (**Figure 14D-F**).

The rapid efficiency enhancement of small-area (<0.1 $cm^2$) series-connected solar cells has primarily been driven by significant progress in the bottom subcell utilizing NBG perovskite materials (approximately 1.25 eV). However, challenges persist for the top subcell employing WBG perovskite materials (>1.75 eV), particularly in the context of large-area (>1 $cm^2$) tandem solar cells. Given this, He et al. introduced a novel approach by developing a SAM of 4-(7H-dibenzo [c, g] carbazole-7-yl) butyl phosphonic acid (4PADCB) as a hole-selective layer for WBG PSCs (**Figure 15A**) [129].The strategy aimed to address the growth of high-quality WBG perovskite films over large areas while mitigating interface nonradiative recombination and facilitating efficient hole extraction (**Figure 15B**). Researchers achieved high-quality, large-area WBG perovskite films (1.77 eV bandgap) by incorporating 4PADCB into the device architecture. This resulted in a high open-circuit voltage ($V_{OC}$) of 1.31 V and a low $V_{OC}$ deficit of 0.46 V. Integrating these subcells led to a monolithic all-perovskite tandem solar cell (1.044 $cm^2$ aperture area) with a certified power conversion efficiency (PCE) of 27.01%, a high $V_{OC}$ of 2.12 V, and an excellent fill factor (FF) of 82.6%.

Self-assembled monolayers (SAMs) can be tailored by modifying their molecular components. While introducing heteroatoms (N, S, and O) into their π-conjugated core is thought to enhance charge transport due to increased electron density, it can also compromise structural integrity by creating reactive sites and potential instability under external conditions. Zhao et al. reported a novel self-assembled molecule named (2-(pyrene-1-yl) ethyl) phosphoric acid (Py3), which, in contrast to conventional molecules such as 2PACz, consists of a polyaromatic core without heteroatom substitution (**Figure 16n**) [130]. The pyrene yields better carrier transport and selectivity while the molecular contacts generated by the pyrene core structure exhibit chemical inertness and structural rigidity to greatly improves the performance of PSCs in terms of efficiency and durability. Furthermore, in different accelerated aging tests (**Figure 16o**), the devices constructed with Py3 as the hole transport layer have an expected T90 (the time it takes for the device to degrade to 90% of its initial efficiency) of more than 10,000 h, which far exceeds the current lifetimes of PSCs in the range of about 3000 h.

### 5.3 Perovskite-silicon tandem solar cells

Crystalline silicon-perovskite tandem solar cell is a photovoltaic technology that combines crystalline silicon solar cells with perovskite solar cells. By superimposing

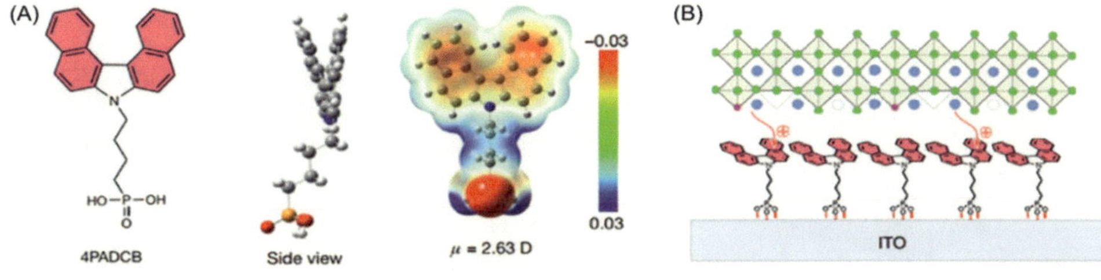


**Figure 15.**
*(A) Molecular structure of 4PADCB and schematic representation of the calculated electrostatic surface potential (ESP) and (B) interconnections with ITO substrates. Reproduced with permission [129]. Copyright 2023,* Nature.

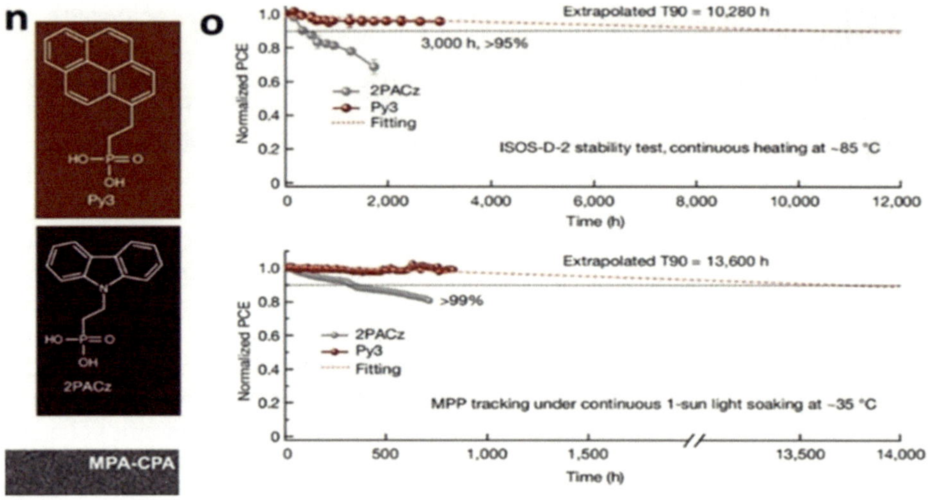


**Figure 16.**
*(n) Molecular structures of 2PACz and Py3. (o) Evolution of the PCEs and the maximum power point (MPP) tracked under different conditions. Reproduced with permission [130]. Copyright 2024,* Nature.

the perovskite cell layer on an efficient crystalline silicon cell to form a series structure, different wavelengths in the solar spectrum can be absorbed more widely, improving the overall photoelectric conversion efficiency. Highly efficient tandem structures include p-i-n and n-i-p configurations based on the charged transport material; either hole or electron is exposed to sunlight first, as illustrated in **Figure 17(a-b)**.

Due to their covalent bonding with the substrate surface, SAMs offer increased durability during perovskite processing and can effectively maintain conformal coverage on textured surfaces. Therefore, SAM-based HTLs are excellent candidates for direct integration into monolithic perovskite-silicon tandem solar cells on textured silicon. Ou et al. successfully demonstrated the excellent stability of MA-free WBG perovskite as a light absorber (**Figure 18A**) [131]. They introduced a novel strategy for simultaneous interface modification and surface passivation to mitigate interface barriers and

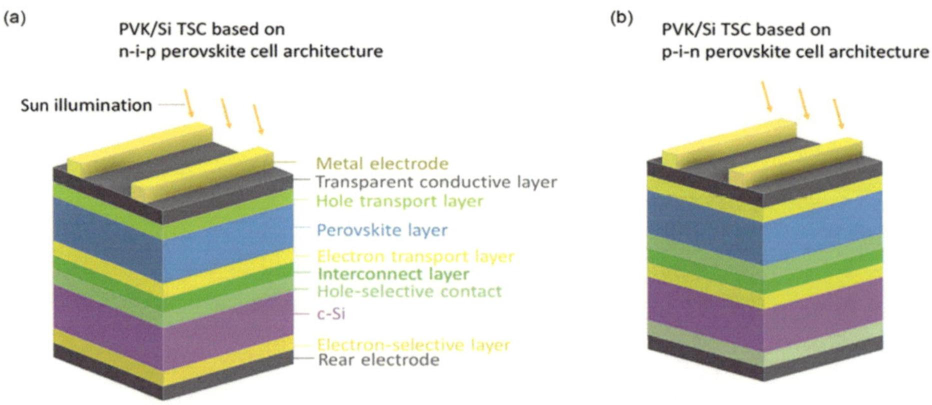


**Figure 17.**
*Schematic of perovskite-Si tandem configurations: (a) n-i-p perovskite-Si TSC architecture and (b) p-i-n perovskite-Si TSC architecture.*

recombination losses in the devices. A mixture of 2PACz and MeO-2PACz was employed as a SAM to reduce interface charge transport losses and promote the growth of larger perovskite grains (**Figure 18B**). A perovskite-Si tandem solar cell with a 4 T architecture was fabricated, achieving an impressive efficiency of 26.59% (**Figure 18C**). Stable and efficient perovskite-Si tandem solar cells require effective defect passivation and suppression of light-induced phase segregation in WBG perovskite materials. Isikgor's research demonstrated the utilization of phenylbiguanide hydrochloride (PhenHCl), a molecule containing both electron-rich and electron-poor parts, to fulfill these requirements independently of the surface chemistry of perovskite, grain boundaries, and interfaces (**Figure 18D**) [132]. When combined with 2PACz SAM as hole-selective contacts for the perovskite layer, PhenHCl passivation further improved the PCE of perovskite-Si tandem solar cells from 25.4 to 27.4%.

To achieve high-performance inverted perovskite-Si tandem cells, Zheng et al. employed 2PACz as the surface modification material to optimize the NiOx/perovskite interface in tungsten-doped indium oxide-based inverted PSCs for tandem devices (**Figure 18E**) [133]. The researchers conducted in-depth investigations into the underlying mechanisms that contributed to the superior performance of the NiOx/SAM HTL, which surpassed the results obtained with either a single NiOx

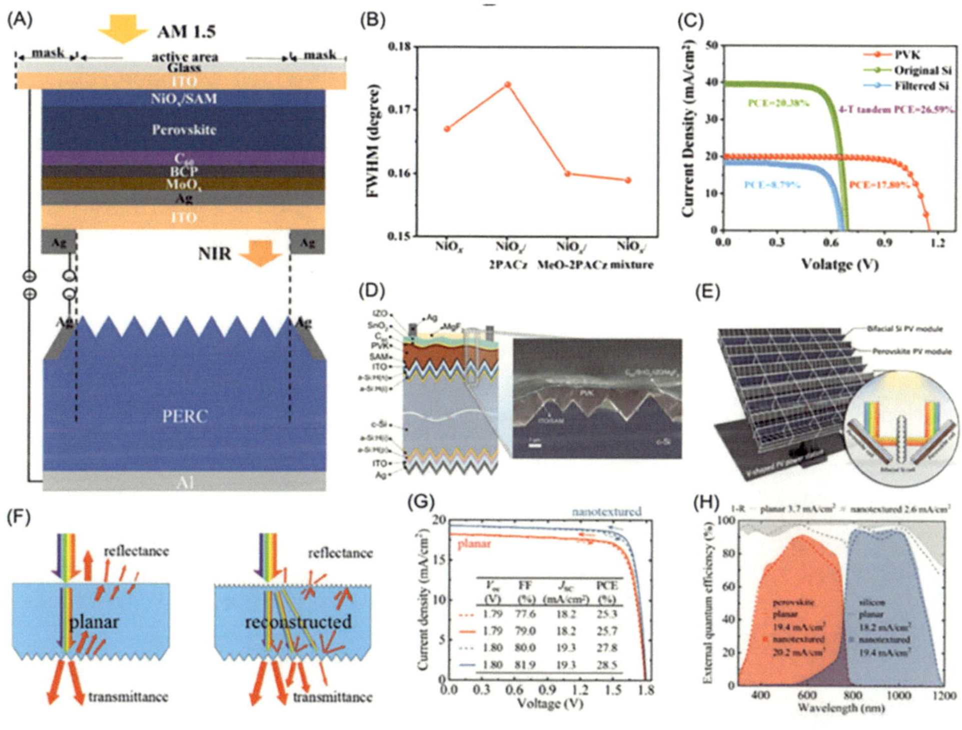


**Figure 18.**
*(A) Diagram of the four-terminal perovskite-Si tandem solar cell framework. Reproduced with permission. (B) Full-width at half-maximum (FWHM) of a perovskite layer in a mixture of 2PACz and MeO-2PACz modified NiOx, and (C) J–V curves of tandem devices based on hybrid SAMs. Reproduced with permission [131]. Copyright 2023,* Chemical Engineering Journal. *(D) Structure of a perovskite-Si tandem solar cell and corresponding cross-sectional images. Reproduced with permission [132]. Copyright 2021,* Joule. *(E) Design of a perovskite-Si tandem device. Reproduced with permission [133]. Copyright 2022,* Journal of Materials Chemistry A. *(F) Light capture mechanism diagrams for planar and reconstructed b-Si samples, (G) J-V curves, and (H) EQE curves for planar and nanostructured perovskite-Si tandem solar cells. Reproduced with permission [134]. Copyright 2022,* Joule.

or SAM HTL. Ying et al. successfully demonstrated a monolithic perovskite/silicon tandem solar cell based on the tunnel oxide passivated contact (TOPCon) structure with nanostructured black silicon (b-Si) subcell (**Figure 17F**) [134]. They introduced a carbazole-based MeO-2PACz SAM as a hole-selective contact at the indium zinc oxide/perovskite interface. These advancements significantly improved the series-connected JSC and FF, ultimately resulting in a PCE of 28.5% (**Figure 18G-H**).

Mishima conducted a study on perovskite-Si tandem solar cells, examining the impact of blending two different SAMs containing carbazole cores (**Figure 19A**) [135]. The 2PACz was chosen as a companion SAM for its superior passivation quality and molecular compatibility with MeO-2PACz (**Figure 19B**). MeO-2PACz showed poor passivation due to spatial effects, leaving ITO surface areas exposed. Blending it with 2PACz, a smaller ligand, improved passivation by covering these gaps. Building on Ph-2PACz, Wang et al. developed a new carbazole SAM with two phenyl rings. (**Figure 19C**) [136]. This new SAM hole-selective layer, featuring an extended conjugated system for high-bandgap perovskite solar cells (PSCs) and tandem structures, has a narrowed bandgap offset that better aligns its HOMO level with the perovskite's VBM. It also improves surface wettability, leading to better perovskite film formation, an optimized interface, and faster charge extraction.

### 5.4 Perovskite-CIGSe tandem solar cells

Copper indium gallium selenide (CIGS) solar cells, a well-established photovoltaic technology, can be used as a viable bottom cell candidate for double-junction tandem solar cells (TSCs). Recently, the PCE of the most advanced 4 T perovskite/CIGS TSCs reached 29.9%, while the highest PCE of 2 T perovskite/CIGS TSC is 24.2%, which develops relatively slowly. In contrast to the leading perovskite/silicon (Si)TSCs in terms of PCE, perovskite/CIGS TSCs exhibit distinctive advantages such as adjustable bandgap, high absorption coefficient, radiation resistance, and can be prepared on flexible substrates.

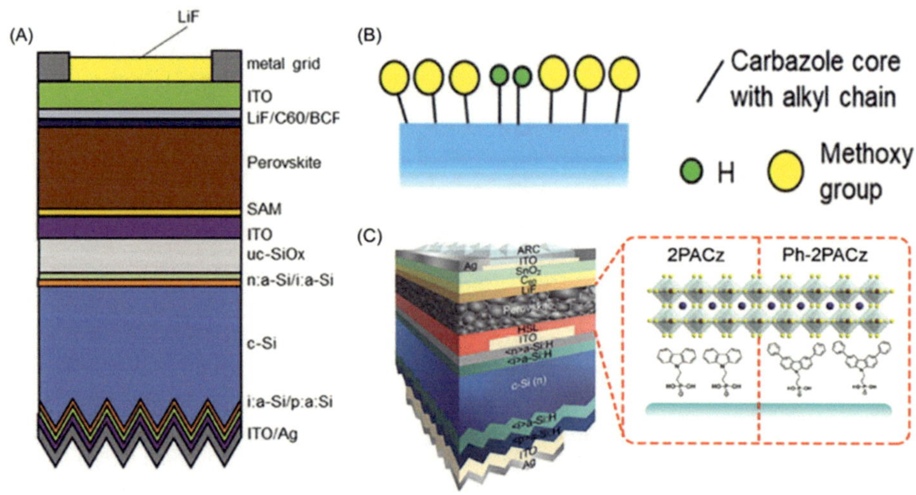


**Figure 19.**
*(A) Structure of a perovskite-Si tandem solar cell based on mixed MeO-2PACz and 2PACz SAMs. (B) Surface property of mixed MeO-2PACz and 2PACz SAMs on ITO substrate. Reproduced with permission [135]. Copyright 2022,* Applied Physics Express. *(C) Schematic of the perovskite/Si tandem solar cell with Ph-2PACz. Reproduced with permission [136]. Copyright 2023,* Joule.

Cu(In, Ga)$Se_2$ (CIGSe) thin-film solar cells possess desirable band gaps, making them a promising choice for integration into perovskite-based tandem solar cells [137]. However, the rough surface of CIGSe presents challenges in the fabrication of thin hole-selective contact layers, which are essential for achieving high-efficiency perovskite top cells [138]. Al-Ashouri et al. have demonstrated the robustness of the self-assembly process even when confronted with the challenges posed by the rugged surface of CIGSe, which was realized through the immersion of a CIGSe bottom cell into a solution containing MeO-2PACz. Through the seamless integration of SAM into a monolithic perovskite/CIGSe tandem solar cell, they successfully prepared a device that consistently maintained a stable efficiency of 23.26% (area of 1.03 $cm^2$) [139].

Kafedjiska et al. conducted a comprehensive investigation into the performance of five different HTLs in monolithic perovskite and CIGSe tandem solar cells (**Figure 20A**). The results revealed that the NiOx-based devices suffered from a substantial deficit in quasi-Fermi level splitting at open circuit (QFLS-$V_{OC}$). The deficit stemmed from issues such as poor electron capture, inefficient hole extraction at the NiOx/perovskite interface, and a shorter effective carrier lifetime within the majority of the perovskite layers (**Figure 20B**) [140]. Incorporating 2% copper into NiOx and passivating its surface with MeO-2PACz effectively suppressed electron capture, enhanced hole extraction, reduced nonradiative recombination, and improved band alignment. Brinkmann et al.'s QFLS analysis highlighted $V_{OC}$ losses related to charge extraction layers. Using MeO-2PACz as the HTL created a dense SAM, leading to

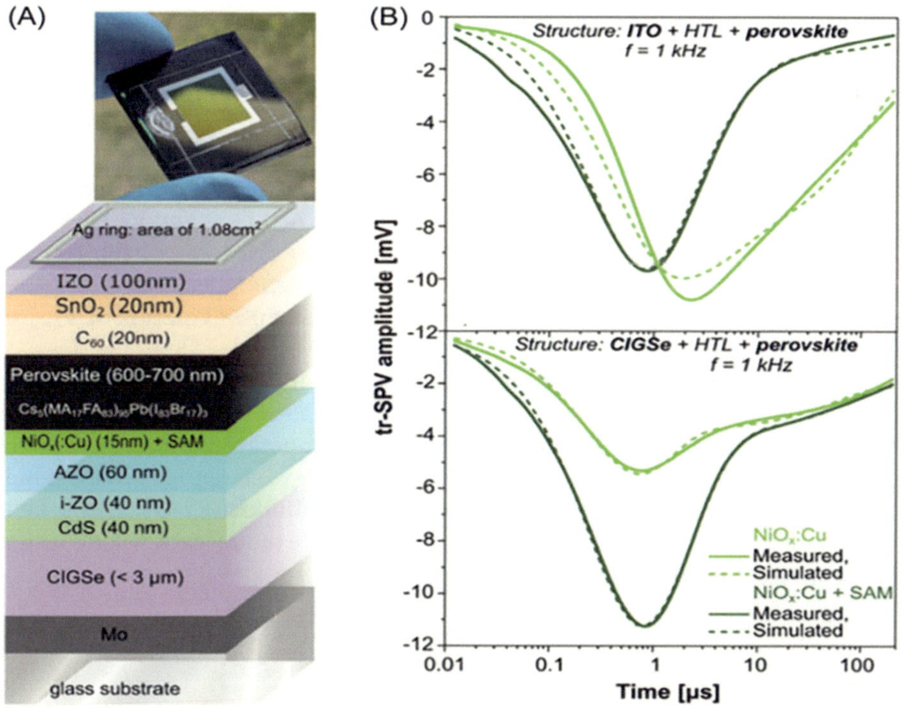


**Figure 20.**
*(A) Structural framework of the tandem perovskite-CIGSe solar cell and (B) tr-SPV measurements at f = 1 kHz based on the tandem device. Reproduced with permission [140]. Copyright 2023,* Advanced Functional Materials.

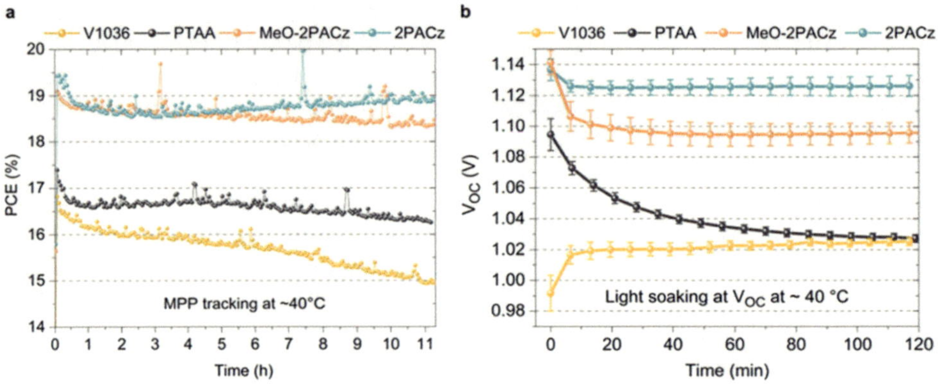


**Figure 21.**
*Stability assessments of PSCs based on the investigated HSCs in $N_2$ atmosphere. (a) MPP tracking under continuous, simulated 1 sun AM 1.5G illumination of uncooled devices (reaching a temperature of B40 1C in operation). (b) Time evolution of $V_{OC}$ values of solar cells kept at open-circuit conditions under light-soaking at full sun illumination at B40 1C cell temperature (no active cooling). The values were extracted from J–V-scans every 6–8 min. The error bars show the standard deviation of these values across the individual cells (4 V1036, 6 PTAA, 3MeO-2PACz, and 10 2PACz cells). Reproduced with permission [139]. Copyright 2019,* Energy Environmental Science.

a significant 90 meV increase in perovskite QFLS compared to PTAA, indicating improved $V_{OC}$ potential [141].

Amran Al-Ashouri et al. synthesized two new SAM-forming molecules, MeO-2PACz and 2PACz, and demonstrated their ability to conformally create a hole-selective layer on a rough surface within a monolithic CIGSe/perovskite tandem solar cell, achieving a stabilized, certified PCE of 23.26%. Solar cells using these SAMs also showed increased stability compared to those using PTAA. **Figure 21a** shows the time evolution of PCE under continuous MPP tracking for the investigated HSCs at simulated 1 sun AM 1.5G illumination without active sample cooling. Operating solar cells reach at least 40°C. After 11 hours, MeO-2PACz and 2PACz-based devices show only slight PCE losses (<3%) [1] compared to PTAA (~6%) and V1036 (~12%). Under harsher light soaking conditions at open-circuit, which creates a high density of charge carriers, a more significant difference in degradation between the HSCs is expected [139]. The most notable differences occur in the time evolution of $V_{OC}$, as displayed in **Figure 21b**.

Under light soaking at open-circuit, PTAA-based solar cells experience a significant 60 mV drop in $V_{OC}$ within 2 hours, whereas SAM-based cells (MeO-2PACz and 2PACz) show virtually stable $V_{OC}$ after an initial temperature-related drop. The study suggests that excess charge carriers lower the energy barrier for ion movement, leading to iodine diffusion and structural damage in the bulkier PTAA. SAMs, being chemically robust and volumetrically negligible electrode modifications, are resistant to both structural damage and ion accumulation.

## 6. Conclusions and future perspectives

While self-assembly presents transformative opportunities for thin-film photovoltaics, it faces limitations in reproducibility, scalability, and material compatibility. Achieving consistent reproducibility is challenging due to the sensitivity of self-assembly processes to minute variations in experimental conditions, hindering reliable device fabrication. Scaling up self-assembly techniques from the laboratory to industrial production is complex, requiring precise control over large areas and in high-throughput environments. Ensuring material compatibility is crucial, as self-assembly methods may have specific requirements, limiting material choices and potentially affecting device performance. Ongoing research is focused on overcoming these barriers by developing more robust and controllable self-assembly methods, exploring continuous and roll-to-roll processing for enhanced scalability, and investigating novel strategies to expand material compatibility.

The future of self-assembly in thin-film photovoltaics is bright, with several key trends poised to reshape the field: the development of zero-waste coating processes to minimize material usage and promote sustainability, the exploration of novel material systems like 2D materials to enhance optoelectronic properties, and the integration of artificial intelligence to optimize the design and control of self-assembly processes. These advancements, coupled with a growing emphasis on interdisciplinary research that fosters collaboration between chemists, materials scientists, physicists, engineers, and computer scientists, hold the promise of driving the creation of next-generation photovoltaic devices with improved efficiency, stability, and cost-effectiveness, ultimately accelerating the transition toward a cleaner and more sustainable energy future.

This chapter has explored the critical role of self-assembly in advancing thin-film photovoltaic technologies, demonstrating its power as a bottom-up fabrication approach to precisely control nanoscale material organization for enhanced light absorption, charge transport, and overall device efficiency in perovskite, organic, and quantum dot solar cells. Key principles, including molecular interactions, templating techniques, and solvent engineering, were shown to facilitate the creation of ordered thin films with superior optoelectronic properties, leading to significant performance improvements, such as efficiencies exceeding 26% in perovskite solar cells through interfacial engineering with self-assembled monolayers (SAMs). While challenges related to reproducibility, scalability, and material compatibility persist, the future outlook for self-assembly in photovoltaics remains highly promising. Key future directions include the development of more sustainable, zero-waste coating processes, the exploration of novel material systems like 2D materials, and the integration of artificial intelligence to optimize self-assembly design and control. Continued interdisciplinary collaboration will be crucial to overcome existing limitations and fully realize the transformative potential of self-assembly, paving the way for next-generation photovoltaic devices with greater efficiency, stability, and cost-effectiveness for a sustainable energy future.

## Acronym

| | |
|---|---|
| SAM | self-assembly materials |
| PSC | perovskite solar cell |
| PCE | power conversion efficiency |
| FT-IR | Fourier transform infrared |
| FF | fill factor |
| VOC | open-circuit voltage |
| JSC | short-circuit current density |
| MPPT | maximum power-point tracking |
| TEM | transmission electron microscopy |

| | |
|---|---|
| NWs | nanowires |
| SA | self-assembly |
| HTL | hole transport layer |
| ETL | electron transport layer |
| FTO | fluorine-doped tin oxide |
| ITO | indium tin oxide |
| NiO | nickel oxide |
| WF | work function |
| PVD | physical vapor deposition |
| SEM | scanning electron microscopy |
| AFM | atomic force microscopy |
| XRD | X-ray diffraction |
| $MAPbI_3$ | methylammonium lead iodide |
| $MA^+$ | methylammonium |
| $FA^+$ | formamidinium |
| DSSC | dye-sensitized solar cell |
| ZTO | zinc tin oxide |
| OPVs | organic photovoltaics |
| LbL | layer-by-layer |
| DIO | 1,8-diiodooctane |
| QDs | quantum dots |
| PCCE | photonic crystal coupled emission |
| CVD | chemical vapor deposition |
| NR | nonradiative |
| TSC | tandem solar cell |
| WBG | wide-bandgap |
| NBG | normal-bandgap |
| LBG | low-bandgap |
| b-Si | black silicon |
| CIGS | copper indium gallium selenide |

## Author details

Zhenyu Zhang[1], Zhu Ma[2*], Zhuowei Du[2], Yi Chen[2], Qian Zhang[2], Hao Du[2], Zhuo Lv[2], Siming Huang[1], Maryam Shahrostami[3], Javad Maleki[3], Narges Yaghoobi Nia[4], Mahmoud Zendehdel[4,5,6*] and Mojtaba Abdi-Jalebi[1*]

1 Institute for Materials Discovery University College London Malet Place, London, UK

2 School of New Energy and Materials, Southwest Petroleum University, Chengdu, China

3 Department of Electrical and Computer Engineering, Tarbiat Modares University (TMU), Tehran, Iran

4 Kimia Solar Research Institute (K.S.R.I), Kimia Solar Company, Kashan, Iran

5 Iritaly Trading Company S.r.l., Via Volturno, Rome, Italy

6 Department of Electronics Engineering, University of Rome Tor Vergata, Roma, Italy

*Address all correspondence to: zhuma@swpu.edu.cn; m.zendehdel@iritalytrading.com and m.jalebi@ucl.ac.uk

## References

[1] Kim JY, Lee JW, Jung HS, Shin H, Park NG. High-efficiency perovskite solar cells. Chemical Reviews. 2020;**120**:7867-7918. DOI: 10.1021/acs.chemrev.0c00107

[2] Chen M et al. Revolutionizing perovskite solar cells with carbon electrodes: Innovations and economic potential. Advanced Functional Materials. 2025;**35**:2422020. DOI: 10.1002/adfm.202422020

[3] Baumann S et al. Stability and reliability of perovskite containing solar cells and modules: Degradation mechanisms and mitigation strategies. Energy & Environmental Science. 2024;**17**:7566-7599. DOI: 10.1039/d4ee01898b

[4] Takhellambam D et al. Enhancing hole transfer in perovskite solar cell with self-assembled monolayer by introducing [1]Benzothieno [3,2-b] [1] Benzothiophene interlayer. Solar RRL. 2023;**7**:2300658. DOI: 10.1002/solr.202300658

[5] Chu L, Ding L. Self-assembled monolayers in perovskite solar cells. Journal of Semiconductors. 2021;**42**:090202. DOI: 10.1088/1674-4926/42/9/090202

[6] Le Corre VM et al. Charge transport layers limiting the efficiency of perovskite solar cells: How to optimize conductivity, doping, and thickness. ACS Applied Energy Materials. 2019;**2**:6280-6287. DOI: 10.1021/acsaem.9b00856

[7] Kim SY, Cho SJ, Byeon SE, He X, Yoon HJ. Self-assembled monolayers as Interface engineering nanomaterials in perovskite solar cells. Advanced Energy Materials. 2020;**10**:2002606. DOI: 10.1002/aenm.202002606

[8] Xu D, Wu P, Tan H. Self-assembled monolayers for perovskite solar cells. Information & Functional Materials. 2024;**1**:2-25. DOI: 10.1002/ifm2.8

[9] Choi K et al. A short review on Interface engineering of perovskite solar cells: A self-assembled monolayer and its roles. Solar RRL. 2019;**4**:1900251. DOI: 10.1002/solr.201900251

[10] Yu M-H et al. Impact of self-assembled monolayer structural design on perovskite phase regulation, hole-selective contact, and energy loss In inverted perovskite solar cells. Nano Energy. 2024;**132**:110405. DOI: 10.1016/j.nanoen.2024.110405

[11] Liu H-R et al. Pyridine-functionalized fullerene electron transport layer for efficient planar perovskite solar cells. ACS Applied Materials & Interfaces. 2019;**11**:23982-23989. DOI: 10.1021/acsami.9b03304

[12] Yang Y et al. Design and applications of hole-selective self-assembled monolayers for perovskite photovoltaics. Chemical Engineering Journal. 2025;**504**:158870. DOI: 10.1016/j.cej.2024.158870

[13] Suo J, Yang B, Bogachuk D, Boschloo G, Hagfeldt A. The dual use of SAM molecules for efficient and stable perovskite solar cells. Advanced Energy Materials. 2024;**15**:2400205. DOI: 10.1002/aenm.202400205

[14] Wu T et al. Self-assembled monolayer hole-selective contact for up-scalable and cost-effective inverted perovskite solar cells. Advanced Functional Materials. 2024;**34**:2316500. DOI: 10.1002/adfm.202316500

[15] Singh N, Tao YT. Effect of surface modification of nickel oxide

hole-transport layer via self-assembled monolayers in perovskite solar cells. Nano Select. 2021;**2**:2390-2399. DOI: 10.1002/nano.202100004

[16] Farag A et al. Evaporated self-assembled monolayer hole transport layers: Lossless interfaces in p-i-n perovskite solar cells. Advanced Energy Materials. 2023;**13**:2203982. DOI: 10.1002/aenm.202203982

[17] Navazani S, Yaghoobi Nia N, Zendehdel M, Shokuhfar A, Di Carlo A. Fabrication of high efficiency, low-temperature planar perovskite solar cells via scalable double-step crystal engineering deposition method fully out of glove box. Solar Energy. 2020;**206**:181-187. DOI: 10.1016/j.solener.2020.05.084

[18] Yaghoobi Nia N, Zendehdel M, Cinà L, Matteocci F, Di Carlo A. A crystal engineering approach for scalable perovskite solar cells and module fabrication: A full out of glove box procedure. Journal of Materials Chemistry A. 2018;**6**:659-671. DOI: 10.1039/C7TA08038G

[19] Lan Z-R, Shao J-Y, Zhong Y-W. Self-assembled monolayers as hole-transporting materials for inverted perovskite solar cells. Molecular Systems Design & Engineering. 2023;**8**:1440-1455. DOI: 10.1039/d3me00144j

[20] Park SM et al. Low-loss contacts on textured substrates for inverted perovskite solar cells. Nature. 2023;**624**:289-294. DOI: 10.1038/s41586-023-06745-7

[21] Li D et al. Co-adsorbed self-assembled monolayer enables high-performance perovskite and organic solar cells. Nature Communications. 2024;**15**:7605. DOI: 10.1038/s41467-024-51760-5

[22] Hussain T et al. Experimental evidence of ion migration in aged inorganic perovskite solar cells using non-destructive RBS depth profiling. Materials Advances. 2022;**3**:7846-7853. DOI: 10.1039/d2ma00199c

[23] Liu L et al. Self-assembled amphiphilic monolayer for efficient and stable wide-bandgap perovskite solar cells. Advanced Energy Materials. 2022;**13**:2202802. DOI: 10.1002/aenm.202202802

[24] Tumen-Ulzii G et al. Hysteresis-less and stable perovskite solar cells with A self-assembled monolayer. Communications Materials. 2020;**1**:31. DOI: 10.1038/s43246-020-0028-z

[25] Hung CM et al. Repairing interfacial defects in self-assembled monolayers for high-efficiency perovskite solar cells and organic photovoltaics through the SAM@pseudo-planar monolayer strategy. Advanced Science (Weinheim, Baden-Württemberg, Germany). 2024;**11**:e2404725. DOI: 10.1002/advs.202404725

[26] Zhao X et al. Regulation of buried Interface through the rapid removal of PbI2·DMSO complex for enhancing light stability of perovskite solar cells. ACS Energy Letters. 2024;**9**:2659-2669. DOI: 10.1021/acsenergylett.4c00386

[27] Magomedov A et al. Self-assembled hole transporting monolayer for highly efficient perovskite solar cells. Advanced Energy Materials. 2018;**8**:1801892. DOI: 10.1002/aenm.201801892

[28] Akin Kara D et al. Enhanced device efficiency and long-term stability via boronic acid-based self-assembled monolayer modification of indium tin oxide in a planar perovskite solar cell. ACS Applied Materials & Interfaces. 2018;**10**:30000-30007. DOI: 10.1021/acsami.8b10445

[29] Chang C-Y et al. Achieving high efficiency and improved stability in large-area ITO-free perovskite solar cells with thiol-functionalized self-assembled monolayers. Journal of Materials Chemistry A. 2016;**4**:7903-7913. DOI: 10.1039/c6ta02581a

[30] Guha S. Thin-film photovoltaics: Buffer against degradation. Nature Energy. 2017;**2**:17057. DOI: 10.1038/nenergy.2017.57

[31] Li M, Liu M, Qi F, Lin FR, Jen AKY. Self-assembled monolayers for interfacial engineering in solution-processed thin-film electronic devices: Design, fabrication, and applications. Chemical Reviews. 2024;**124**:2138-2204. DOI: 10.1021/acs.chemrev.3c00396

[32] Li C et al. Self-assembly of block copolymers towards mesoporous materials for energy storage and conversion systems. Chemical Society Reviews. 2020;**49**:4681-4736. DOI: 10.1039/d0cs00021c

[33] Abelson A et al. Collective Topo-epitaxy in the self-assembly of A 3D quantum dot superlattice. Nature Materials. 2019;**19**:49-55. DOI: 10.1038/s41563-019-0485-2

[34] Liang Z et al. Fabrication of tunable aluminum nanodisk arrays via A self-assembly nanoparticle template method and their applications for performance enhancement in organic photovoltaics. Journal of Materials Chemistry A. 2018;**6**:3649-3658. DOI: 10.1039/c7ta10017e

[35] Wolff CM et al. Perfluorinated self-assembled monolayers enhance the stability and efficiency of inverted perovskite solar cells. ACS Nano. 2020;**14**:1445-1456. DOI: 10.1021/acsnano.9b03268

[36] Li Y et al. Self-assembly of chiroptical ionic co-crystals from silver nanoclusters and organic macrocycles. Nature Chemistry. 2025;**17**:169-176. DOI: 10.1038/s41557-024-01696-6

[37] Zuo L et al. Enhanced photovoltaic performance of CH3NH3PbI3 perovskite solar cells through interfacial engineering using self-assembling monolayer. Journal of the American Chemical Society. 2015;**137**:2674-2679. DOI: 10.1021/ja512518r

[38] Yaghoobi Nia N, Matteocci F, Cina L, Di Carlo A. High-efficiency perovskite solar cell based on poly (3-hexylthiophene): Influence of molecular weight and mesoscopic Scaffold layer. ChemSusChem. 2017;**10**:3854-3860. DOI: 10.1002/cssc.201700635

[39] Yaghoobi Nia N et al. Doping strategy for efficient and stable triple cation hybrid perovskite solar cells and module based on poly(3-hexylthiophene) hole transport layer. Small. 2019;**15**:1904399. DOI: 10.1002/smll.201904399

[40] Yang Q et al. Solvent-actuated self-assembly of amphiphilic hole-transporting polymer enables bottom-surface passivation of perovskite film for efficient photovoltaics. Advanced Energy Materials. 2021;**11**:2100493. DOI: 10.1002/aenm.202100493

[41] Zendehdel M, Habibi MH, Hashemzadeh-Esfarjani MS. High efficiency photocatalytic mineralization of C.I. Reactive red 2 using transparent self-assembled nanopin-like titania with high preferential (101) orientation on alumina thin film in aqueous environments. Separation Science and Technology. 2014;**49**:2951-2959. DOI: 10.1080/01496395.2014.945186

[42] Habibi MH, Zendehdel M. Fabrication and characterization of

self-assembled multilayer nanostructure titania with high preferential (101) orientation on alumina thin films by layer by-layer dip-coating method. Current Nanoscience. 2010;**6**:642-647. DOI: 10.2174/157341310793348614

[43] Guo C et al. Light-induced Quinone conformation of polymer donors toward 19.9% efficiency organic solar cells. Energy & Environmental Science. 2024;**17**:2492-2499. DOI: 10.1039/d4ee00605d

[44] Tong X et al. Large orientation angle buried substrate enables efficient flexible perovskite solar cells and modules. Advanced Materials. 2024;**36**:2407032. DOI: 10.1002/adma.202407032

[45] He D et al. Decoupling nucleation and growth of crystal for DMSO-free self-assembly tin perovskite films and their optoelectronic application. Advanced Functional Materials. 2025;**35**:2420593. DOI: 10.1002/adfm.202420593

[46] Guo W et al. Self-assembled molecules fostering ordered spatial heterogeneity for efficient Ruddlesden-popper perovskite solar cells. Advanced Energy Materials. 2024;**14**:2401303. DOI: 10.1002/aenm.202401303

[47] Grzelczak M, Vermant J, Furst EM, Liz-Marzán LM. Directed self-assembly of nanoparticles. ACS Nano. 2010;**4**:3591-3605. DOI: 10.1021/nn100869j

[48] Habibi MH, Mardani M, Habibi M, Zendehdel M. Enhanced photovoltage (Voc) of nano-structured zinc tin oxide (ZTO) working electrode prepared by A green hydrothermal route for dye-sensitized solar cell (DSSC). Journal of Materials Science: Materials in Electronics. 2017;**28**:3789-3795. DOI: 10.1007/s10854-016-5989-9

[49] Habibi MH, Karimi B, Zendehdel M, Habibi M. Preparation of nanostructure mixed copper-zinc oxide via co-precipitation rout for dye-sensitized solar cells: The influence of blocking layer and Co(II)/Co(III) complex redox shuttle. Journal of Industrial and Engineering Chemistry. 2014;**20**:1462-1467. DOI: 10.1016/j.jiec.2013.07.032

[50] Karimian D, Yadollahi B, Zendehdel M, Mirkhani V. Efficient dye-sensitized solar cell with a pure thin film of A hybrid polyoxometalate covalently attached organic dye as a working electrode in a cobalt redox mediator system. RSC Advances. 2015;**5**:76875-76882. DOI: 10.1039/C5RA09104G

[51] Hossein Habibi M, Askari E, Habibi M, Zendehdel MN, Zirconate NZ. Zinc oxide or zirconium oxide pastes coated on fluorine doped tin oxide thin film as photoelectrochemical working electrodes for dye-sensitized solar cell. Spectrochimica Acta Part A: Molecular and Biomolecular Spectroscopy. 2013;**104**:197-202. DOI: 10.1016/j.saa.2012.11.055

[52] Habibi MH, Mikhak M, Zendehdel M, Habibi M. Influence of nanostructured zinc titanate, zinc oxide or titanium dioxide thin film coated on fluorine doped tin oxide as working electrodes for dye-sensitized solar cell. International Journal of Electrochemical Science. 2012;**7**:6787-6798. DOI: 10.1016/S1452-3981(23)15747-6

[53] Habibi MH, Habibi AH, Zendehdel M, Habibi M. Dye-sensitized solar cell characteristics of nanocomposite zinc ferrite working electrode: Effect of composite precursors and titania as a blocking layer on photovoltaic performance. Spectrochimica Acta Part A: Molecular and Biomolecular Spectroscopy. 2013;**110**:226-232. DOI: 10.1016/j.saa.2013.03.051

[54] Habibi MH, Karimi B, Zendehdel M, Habibi M. Fabrication, characterization of two nano-composite Cuo–Zno working electrodes for dye-sensitized solar cell. Spectrochimica Acta Part A: Molecular and Biomolecular Spectroscopy. 2013;**116**:374-380. DOI: 10.1016/j.saa.2013.07.046

[55] Yaghoobi, Nia N, Farahani P, Sabzyan H, Zendehdel M, Oftadeh M. A combined computational and experimental study of the $[Co(Bpy)_3]^{2+/3+}$ complexes as one-electron outer-sphere redox couples in dye-sensitized solar cell electrolyte media. Physical Chemistry Chemical Physics. 2014;**16**:11481-11491. DOI: 10.1039/C3CP55034F

[56] Liao Q et al. Self-assembled donor-acceptor hole contacts for inverted perovskite solar cells with an efficiency approaching 22%: The impact of anchoring groups. Journal of Energy Chemistry. 2022;**68**:87-95. DOI: 10.1016/j.jechem.2021.11.001

[57] Zhu J et al. A donor–acceptor-type hole-selective contact reducing non-radiative recombination losses in both subcells towards efficient all-perovskite tandems. Nature Energy. 2023;**8**:714-724. DOI: 10.1038/s41560-023-01274-z

[58] Desiraju GR. Supramolecular synthons in crystal engineering—A new organic synthesis. Angewandte Chemie International Edition in English. 1995;**34**:2311-2327. DOI: 10.1002/anie.199523111

[59] Moulton B, Zaworotko MJ. From molecules to crystal engineering: Supramolecular isomerism and polymorphism in network solids. Chemical Reviews. 2001;**101**:1629-1658. DOI: 10.1021/cr9900432

[60] Lehn J-M. Perspectives in supramolecular chemistry—From molecular recognition towards molecular information processing and self-organization. Angewandte Chemie International Edition. 1990;**29**:1304-1319. DOI: 10.1002/anie.199013041

[61] Whang D, Jin S, Wu Y, Lieber CM. Large-scale hierarchical Organization of Nanowire Arrays for integrated nanosystems. Nano Letters. 2003;**3**:1255-1259. DOI: 10.1021/nl0345062

[62] Park S-J, Lazarides AA, Mirkin CA, Letsinger RL. Directed assembly of periodic materials from protein and oligonucleotide-modified nanoparticle building blocks. Angewandte Chemie International Edition. 2001;**40**:2909-2912. DOI: 10.1002/1521-3773 (20010803)40: 15<2909::AID- ANIE2909>3.0.CO;2-O

[63] Grzybowski BA, Radkowski M, Campbell CJ, Lee JN, Whitesides GM. Self-assembling fluidic machines. Applied Physics Letters. 2004;**84**:1798-1800. DOI: 10.1063/1.1664019

[64] Yan M, Zhang HT, Widjaja EJ, Chang RPH. Self-assembly of well-aligned gallium-doped zinc oxide Nanorods. Journal of Applied Physics. 2003;**94**:5240-5246. DOI: 10.1063/1.1608473

[65] Ottino JM, Wiggins SR, Campbell CJ, Grzybowski BA. Microfluidic mixers: From microfabricated to self-assembling devices. Philosophical Transactions of the Royal Society of London. Series A: Mathematical, Physical and Engineering Sciences. 2004;**362**:1069-1086. DOI: 10.1098/rsta.2003.1363

[66] Fialkowski M, Bitner A, Grzybowski BA. Self-assembly of polymeric microspheres of complex internal structures. Nature Materials. 2005;**4**:93-97. DOI: 10.1038/nmat1267

[67] Clark TD, Ferrigno R, Tien J, Paul KE, Whitesides GM. Template-directed self-assembly of 10-μm-sized hexagonal plates. Journal of the American Chemical Society. 2002;**124**:5419-5426. DOI: 10.1021/ja020056o

[68] Clark TD, Tien J, Duffy DC, Paul KE, Whitesides GM. Self-assembly of 10-μm-sized objects into ordered three-dimensional arrays. Journal of the American Chemical Society. 2001;**123**:7677-7682. DOI: 10.1021/ja010634l

[69] Gracias DH, Tien J, Breen TL, Hsu C, Whitesides GM. Forming electrical networks in three dimensions by self-assembly. Science. 2000;**289**:1170-1172. DOI: 10.1126/science.289.5482.1170

[70] Boncheva M et al. Magnetic self-assembly of three-dimensional surfaces from planar sheets. Proceedings of the National Academy of Sciences. 2005;**102**:3924-3929. DOI: 10.1073/pnas.0500807102

[71] Yeh HJJ, Smith JS. Fluidic self-assembly for the integration of Gaas light-emitting diodes on Si substrates. IEEE Photonics Technology Letters. 1994;**6**:706-708. DOI: 10.1109/68.300169

[72] Anggi RA et al. Surface modification of $TiO_2$ nanoparticles with terephthalic acid in supercritical carbon dioxide. The Journal of Supercritical Fluids. 2021;**174**:105245. DOI: 10.1016/j.supflu.2021.105245

[73] Bishop KJM, Wilmer CE, Soh S, Grzybowski BA. Nanoscale forces and their uses in self-assembly. Small. 2009;**5**:1600-1630. DOI: 10.1002/smll.200900358

[74] Wagner JP, Schreiner PR. London dispersion in molecular chemistry—Reconsidering steric effects. Angewandte Chemie International Edition. 2015;**54**:12274-12296. DOI: 10.1002/anie.201503476

[75] Battezzati M, Magnasco V. Interaction energy of coupled permanent dipoles in oriented molecules. Chemical Physics Letters. 2010;**496**:362-364. DOI: 10.1016/j.cplett.2010.07.064

[76] Ji Y, Wang M, Yang Z, Ji S, Qiu H. Nanowire-assisted self-assembly of one-dimensional nanocrystal superlattice chains. Journal of Materials Chemistry C. 2019;**7**:8471-8476. DOI: 10.1039/C9TC02118C

[77] Evans CG, O'Brien J, Winfree E, Murugan A. Pattern recognition in the nucleation kinetics of non-equilibrium self-assembly. Nature. 2024;**625**:500-507. DOI: 10.1038/s41586-023-06890-z

[78] Abarzhi SI, Goddard III WA. Interfaces and mixing: Nonequilibrium transport across the scales. Proceedings of the National Academy of Sciences of the United States of America. 2019;**116**:18171-18174. DOI: 10.1073/pnas.1818855116

[79] Bell JB, Day MS, Rendleman CA, Woosley SE, Zingale M. Direct numerical simulations of type Ia supernovae flames. II. The Rayleigh-Taylor instability. The Astrophysical Journal. 2004;**608**:883. DOI: 10.1086/420841

[80] Seo K, Kim M, Ahn JK, et al. Effects of drop size and measuring condition on static contact angle measurement on a superhydrophobic surface with goniometric technique. Korean Journal of Chemical Engineering. 2015;**32**:2394-2399. DOI: 10.1007/s11814-015-0034-x

[81] Sampaio PGV, González MOA. A review on organic photovoltaic cell. International Journal of Energy Research. 2022;**46**:17813-17828. DOI: 10.1002/er.8456

[82] Ghosh T, Panicker JS, Nair VC. Self-assembled organic materials for photovoltaic application. Polymers (Basel). 2017;**9**:112. DOI: 10.3390/polym9030112

[83] Chen M-C, Chiou Y-S, Chiu J-M, Tedla A, Tai Y. Marked improvement in the stability of small molecule organic photovoltaics by interfacial modification using self-assembled monolayers to prevent indium diffusion into the active layer. Journal of Materials Chemistry A. 2013;**1**:3680-3687. DOI: 10.1039/c3ta01229h

[84] Zhao L et al. A review on smart strategies for active layer phase separation regulation of organic solar cells. APL Materials. 2023;**11**:120601. DOI: 10.1063/5.0181582

[85] Cai Y et al. Vertically optimized phase separation with improved exciton diffusion enables efficient organic solar cells with thick active layers. Nature Communications. 2022;**13**:2369. DOI: 10.1038/s41467-022-29803-6

[86] Kim M et al. Critical factors governing vertical phase separation in polymer–PCBM blend films for organic solar cells. Journal of Materials Chemistry A. 2016;**4**:15522-15535. DOI: 10.1039/c6ta06508b

[87] Zhang L et al. Vertical phase separation in bulk heterojunction solar cells formed by in situ polymerization of Fulleride. Scientific Reports. 2014;**4**:5071. DOI: 10.1038/srep05071

[88] Yun J et al. Light trapping in bendable organic solar cells using silica nanoparticle arrays. Energy & Environmental Science. 2015;**8**:932-940. DOI: 10.1039/c4ee01100g

[89] Ju S et al. Dielectric light-trapping nanostructure for enhanced light absorption in organic solar cells. Scientific Reports. 2023;**13**:20649. DOI: 10.1038/s41598-023-47898-9

[90] Firdaus Y et al. Long-range Exciton diffusion in molecular non-fullerene acceptors. Nature Communications. 2020;**11**:5220. DOI: 10.1038/s41467-020-19029-9

[91] Mikhnenko OV, Blom PWM, Nguyen T-Q. Exciton diffusion in organic semiconductors. Energy & Environmental Science. 2015;**8**:1867-1888. DOI: 10.1039/C5EE00925A

[92] Menke SM, Holmes RJ. Exciton diffusion in organic photovoltaic cells. Energy & Environmental Science. 2014;**7**:499-512. DOI: 10.1039/c3ee42444h

[93] Hsu P-J, Tedla A, Zharnikov M, Tai Y. Dual interfacial modifications of an organic solar cell by self-assembled monolayers. Journal of Photochemistry and Photobiology A: Chemistry. 2022;**422**:113554. DOI: 10.1016/j.jphotochem.2021.113554

[94] Ruiz-Carretero A et al. Stepwise self-assembly to improve solar cell morphology. Journal of Materials Chemistry A. 2013;**1**:11674-11681. DOI: 10.1039/c3ta12411h

[95] Haruk AM, Mativetsky JM. Supramolecular approaches to nanoscale morphological control in organic solar cells. International Journal of Molecular Sciences. 2015;**16**:13381-13406. DOI: 10.3390/ijms160613381

[96] Kang SJ et al. Using self-organization to control morphology in molecular photovoltaics. Journal of the American Chemical Society. 2013;**135**:2207-2212. DOI: 10.1021/ja308628z

[97] Zhang R et al. Equally high efficiencies of organic solar cells

processed from different solvents reveal key factors for morphology control. Nature Energy. 2024;**10**:124-134. DOI: 10.1038/s41560-024-01678-5

[98] Lee KC, Kim S, Son J, Kong J, Yang C. A brief review on self-assembled monolayers in organic solar cells: Progress, challenges, and future prospects. ACS Applied Electronic Materials. 2025;**7**:946-963. DOI: 10.1021/acsaelm.4c01922

[99] Yip HL, Hau SK, Baek NS, Ma H, Jen AKY. Polymer solar cells that use self-assembled-monolayer-modified ZnO/metals as cathodes. Advanced Materials. 2008;**20**:2376-2382. DOI: 10.1002/adma.200703050

[100] Heo J et al. Electro-inductive effect: Electrodes as functional groups with tunable electronic properties. Science. 2020;**370**:214-219. DOI: 10.1126/science.abb6375

[101] Beard MC. Multiple exciton generation in semiconductor quantum dots. The Journal of Physical Chemistry Letters. 2011;**2**:1282-1288. DOI: 10.1021/jz200166y

[102] Hanna MC, Nozik AJ. Solar conversion efficiency of photovoltaic and photoelectrolysis cells with carrier multiplication absorbers. Journal of Applied Physics. 2006;**100**:074510. DOI: 10.1063/1.2356795

[103] Riel BJ. An introduction to self-assembled quantum dots. American Journal of Physics. 2008;**76**:750-757. DOI: 10.1119/1.2907856

[104] Jiao M et al. Strongly emitting and long-lived silver indium Sulfide quantum dots for bioimaging: Insight into co-ligand effect on enhanced photoluminescence. Journal of Colloid and Interface Science. 2020;**565**:35-42. DOI: 10.1016/j.jcis.2020.01.006

[105] Kim G-H et al. High-efficiency colloidal quantum dot photovoltaics via robust self-assembled monolayers. Nano Letters. 2015;**15**:7691-7696. DOI: 10.1021/acs.nanolett.5b03677

[106] Yuan S et al. Self-assembled high quality CsPbBr3 quantum dot films toward highly efficient light-emitting diodes. ACS Nano. 2018;**12**:9541-9548. DOI: 10.1021/acsnano.8b05185

[107] Majumdar A, Rundquist A, Bajcsy M. Design and analysis of photonic crystal coupled cavity arrays for quantum simulation. Physical Review B. 2012;**86**:195312. DOI: 10.1103/PhysRevB.86.195312

[108] García Ruiz JM et al. Morphogenesis of self-assembled nanocrystalline materials of barium carbonate and silica. Science. 2009;**323**:362-365. DOI: 10.1126/science.1165349

[109] Fan Z, Grünwald M. Orientational order in self-assembled nanocrystal superlattices. Journal of the American Chemical Society. 2019;**141**:1980-1988. DOI: 10.1021/jacs.8b10752

[110] Campos RD et al. Self-assembled molecules for hole extraction in efficient inverted Pbs quantum dot solar cells. Journal of Materials Chemistry A. 2024;**12**:32689-32696. DOI: 10.1039/d4ta04791e

[111] Jernigan GG, Thompson PE. Scanning tunneling microscopy of Sige alloy surfaces grown on Si(100) by molecular beam epitaxy. Surface Science. 2002;**516**:207-215. DOI: 10.1016/s0039-6028(02)02027-7

[112] Sautter KE, Vallejo KD, Simmonds PJ. Strain-driven quantum dot self-assembly by molecular beam epitaxy. Journal of Applied Physics. 2020;**128**:031101. DOI: 10.1063/5.0012066

[113] Vallejo KD, Cabrera-Perdomo CI, Garrett TA, Drake MD, Liang B, Grossklaus KA, et al. Tunable mid-infrared interband emission from tensile-strained InGaAs quantum dots. ACS Nano. 2023;**17**:2318-2327. DOI: 10.1021/acsnano.2c08985

[114] Nagayama K. Two-dimensional self-assembly of colloids in thin liquid films. Colloids and Surfaces A: Physicochemical and Engineering Aspects. 1996;**109**:363-374. DOI: 10.1016/0927-7757(95)03467-6

[115] Mukai K, Ishida Y. Improvement of solar cell characteristics using PbS quantum dot superlattice prepared by sedimentation. Journal of Nanomaterials. 2018;**2018**:3617687. DOI: 10.1155/2018/3617687

[116] Emanuele Marino AS, Berkhout A, MacArthur KE, Heggen M, Gregorkiewicz T, Kodger TE, et al. Simultaneous photonic and excitonic coupling in spherical quantum dot supercrystals. ACS Nano. 2020;**14**:13806-13815. DOI: 10.1021/acsnano.0c06188

[117] Siyu Zheng YW, Jia D, Tian L, Chen J, Shan L, Dong L, et al. Strong coupling of colloidal quantum dots via self-assemble passivation for efficient infrared solar cells. Advanced Materials Interfaces. 2021;**8**:2100489. DOI: 10.1002/admi.202100489

[118] Igor Coropceanu EMJ, Portner J, Haubold D, Nguyen TD, Das A, Tanner CPN, et al. Self-assembly of nanocrystals into strongly electronically coupled all-inorganic supercrystals. Science. 2022;**375**:1422-1426. DOI: 10.1126/science.abm6753

[119] Ma W et al. Photovoltaic performance of ultrasmall Pbse quantum dots. ACS Nano. 2011;**5**:8140-8147. DOI: 10.1021/nn202786g

[120] Pattantyus-Abraham AG et al. Depleted-heterojunction colloidal quantum dot solar cells. ACS Nano. 2010;**4**:3374-3380. DOI: 10.1021/nn100335g

[121] Piliego C, Protesescu L, Bisri S, Kovalenko M, Loi M. 5.2% efficient Pbs nanocrystal Schottky solar cells. Energy & Environmental Science. 2013;**6**:3054-3059. DOI: 10.1039/c3ee41479e

[122] Shi G et al. Finely interpenetrating bulk heterojunction structure for Lead Sulfide colloidal quantum dot solar cells by convective assembly. ACS Energy Letters. 2019;**4**:960-967. DOI: 10.1021/acsenergylett.9b00053

[123] Semichaevsky AV, Johnson HT. Carrier transport in A quantum dot solar cell using Semiclassical and quantum mechanical models. Solar Energy Materials & Solar Cells. 2013;**108**:189-199. DOI: 10.1016/j.solmat.2012.09.030

[124] Norman AG et al. InGaAs/GaAs QD Superlattices - MOVPE growth, structural and optical characterization, and application in intermediate-band solar cells. In: Conference Record of the IEEE Photovoltaic Specialists Conference, 31st. IEEE; 2006. pp. 43-48. DOI: 10.1109/PVSC.2005.1488065

[125] Yi Z et al. Self-assembled monolayers (SAMs) in inverted perovskite solar cells and their tandem photovoltaics application. Interdisciplinary Materials. 2024;**3**:203-244. DOI: 10.1002/idm2.12145

[126] Thiesbrummel J et al. Understanding and minimizing

VOC losses in all-perovskite tandem photovoltaics. Advanced Energy Materials. 2023;**13**:2202674. DOI: 10.1002/aenm.202202674

[127] Lai H et al. High-performance flexible all-perovskite tandem solar cells with reduced VOC-deficit In wide-bandgap subcell. Advanced Energy Materials. 2022;**12**:2202438. DOI: 10.1002/aenm.202202438

[128] Wang W et al. Versatile self-assembled molecule enables high-efficiency wide-bandgap perovskite solar cells and organic solar cells. Advanced Energy Materials. 2023;**13**:2300694. DOI: 10.1002/aenm.202300694

[129] He R et al. Improving Interface quality for 1-cm2 all-perovskite tandem solar cells. Nature. 2023;**618**:80-86. DOI: 10.1038/s41586-023-05992-y

[130] Zhao K et al. Peri-fused Polyaromatic molecular contacts for perovskite solar cells. Nature. 2024;**632**:301-306. DOI: 10.1038/s41586-024-07712-6

[131] Ou Y et al. Collaborative interfacial modification and surficial passivation for high-efficiency MA-free wide-bandgap perovskite solar cells. Chemical Engineering Journal. 2023;**469**:143860. DOI: 10.1016/j.cej.2023.143860

[132] Isikgor FH et al. Concurrent cationic and anionic perovskite defect passivation enables 27.4% perovskite/silicon tandems with suppression of halide segregation. Joule. 2021;**5**:1566-1586. DOI: 10.1016/j.joule.2021.05.013

[133] Zheng L et al. Inverted perovskite/silicon V-shaped tandem solar cells with 27.6% efficiency via self-assembled monolayer-modified nickel oxide layer. Journal of Materials Chemistry A. 2022;**10**:7251-7262. DOI: 10.1039/D1TA10313J

[134] Ying Z et al. Monolithic perovskite/black-silicon tandems based on tunnel oxide passivated contacts. Joule. 2022;**6**:2644-2661. DOI: 10.1016/j.joule.2022.09.006

[135] Mishima R et al. 28.3% efficient perovskite-silicon tandem solar cells with mixed self-assembled monolayers. Applied Physics Express. 2022;**15**:076503. DOI: 10.35848/1882-0786/ac727b

[136] Wang G et al. Molecular engineering of hole-selective layer for high band gap perovskites for highly efficient and stable perovskite-silicon tandem solar cells. Joule. 2023;7:2583-2594. DOI: 10.1016/j.joule.2023.09.007

[137] Jost M et al. Perovskite/CIGS tandem solar cells: From certified 24.2% toward 30% and beyond. ACS Energy Letters. 2022;7:1298-1307. DOI: 10.1021/acsenergylett.2c00274

[138] Liu X et al. Over 28% efficiency perovskite/Cu (InGa) $Se_2$ tandem solar cells: Highly efficient sub-cells and their bandgap matching. Energy Environmental Science. 2023;**16**:5029-5042. DOI: 10.1039/D3EE00869J

[139] Al-Ashouri A et al. Conformal monolayer contacts with lossless interfaces for perovskite single junction and monolithic tandem solar cells. Energy environmental. Science. 2019;**12**:3356-3369. DOI: 10.1039/C9EE02268F

[140] Kafedjiska I et al. Advanced characterization and optimization of NiOx: Cu-SAM hole-transporting Bi-layer for 23.4% efficient monolithic

Cu (In, Ga) $Se_2$-perovskite tandem solar cells. Advanced Functional Materials. 2023;**33**:2302924. DOI: 10.1002/adfm.202302924

[141] Brinkmann K et al. Perovskite–organic tandem solar cells with indium oxide interconnect. Nature. 2022;**604**:280-286. DOI: 10.1038/s41586-022-04455-0

Chapter 4

# Preparation and Magnetic Properties of $Fe_3O_4$@rGO Nanocomposite Investigated by ESR Spectroscopy

*Mondher Yahya and Haythem Barrak*

## Abstract

In this chapter, we present an experimental case study concerning a nanocomposite based on $Fe_3O_4$ and graphene which was synthesized by a simple method. This class of $Fe_3O_4$-carbon hybrid nanocomposites has been tested for various applications. Graphene oxide (GO) has attracted great interest in nanotechnology due to its interesting properties, such as mechanical strength, high electron transport capacity a large specific surface area, thermal conductivity, chemical modifiability, elasticity, and tunable optical properties. Additionally, the functional groups present on GO (-OH, -COOH, and -CHO) have made the grafting of multiple organic compounds uncomplicated. Recently, $Fe_3O_4$ has been implanted in a non-magnetic graphene matrix to combine the properties of graphene with the magnetic properties of nanomaterials. The combination of graphene oxide with $Fe_3O_4$ forms $Fe_3O_4$@rGO nanocomposites. We used these nanocomposites for the extraction of rare earth elements from a phosphoric acid medium, owing to their excellent anticorrosion properties and facile magnetic recovery, and for their application as aqueous electrochemical supercapacitors. The magnetic properties of $Fe_3O_4$ NPs in $Fe_3O_4$@rGO composites were investigated using electron spin resonance spectroscopy (ESR).

**Keywords:** nanocomposites, $Fe_3O_4$@rGO, ferromagnetic state, ESR, rare earth extraction, supercapacitor

## 1. Introduction

Iron played a crucial role in the evolution of civilizations and technologies. Synonymous with progress, and inseparable from the greatest technological achievements, iron has no real rivals in the different fields of application. Iron can, depending on the temperature and the equilibrium oxygen pressure, give rise to a multitude of oxides. Iron oxides have captivated scientific interest for centuries, not only for their ubiquity in nature but also for their remarkable versatility in technological applications. This chapter delves into the fascinating world of magnetite ($Fe_3O_4$), a versatile iron oxide. We explore the intricacies of its synthesis and characterization; with a particular emphasis on unlocking the secrets behind its magnetic properties. Beyond their natural occurrence in the Earth's crust, a range of chemical methods meticulously craft iron oxides with tailored properties for industrial use are indispensable.

Iron oxides have dielectric, magnetic, and optical properties as well as chemical activity which give them a determining role in technological applications, such as catalysis, pollutant adsorbents, magnetic recording media, microelectronics, lithium-ion batteries, biomedical devices, and as a composite with different polymer matrices [1, 2]. Knowledge of the structural and electronic properties of these oxides is a very important step for the production of more efficient and reliable materials.

Wüstite FeO it is a black-colored powder crystallized in NaCl structure with $Fe^{2+}$ ions. The crystals are always non-stoichiometric and iron deficient. Hematite, chemically known as α-Fe2O3, is the most stable iron oxide under ambient conditions. It has a characteristic reddish-brown tint, the iron atoms are $Fe^{3+}$ ions. These two oxides are insulators and antiferromagnets. Above the Morin TM temperature (260 K) and below Neel temperature TN (950°C), hematite has an uncompensated spin (canted spins); this produces a weak ferromagnetic state [3].

Maghemite ($\gamma$-$Fe_2O_3$) arises from the thermal dehydration of goethite (FeO(OH)) or the oxidation of magnetite, it is weakly ferromagnetic. It has the same chemical formula as hematite and has a crystal structure similar to magnetite [4].

Magnetite ($Fe_3O_4$) is a ferromagnetic iron oxide and half-metal, having a very high Curie temperature ($T_C$ = 860 K). It has an inverse spinel structure, the unit cell is written as follows: $(Fe^{3+})_8[Fe^{3+}Fe^{2+}]_8O_{32}$, where parentheses denote tetrahedral site cations and the hooks denote octahedral cations. The ferrimagnetism originated from the magnetic moment of anti-parallel spins between $Fe^{3+}$ ions at tetrahedral sites and both $Fe^{3+}$ and $Fe^{2+}$ ions at octahedral sites. The half-metallicity is a direct consequence of the double-exchange interaction; the transport is thus carried out by jumping electrons between the $Fe^{2+}$ and $Fe^{3+}$ ions of octahedral sites. The double exchange is a ferromagnetic coupling and necessarily implies that the spins of the cations of octahedral sites are parallel [5, 6]. $Fe_3O_4$ is widely used in many applications due to its low cost, low toxicity, and good magnetic properties. The preparation of magnetite is of great interest to improve its efficiency in various applications and reduce production costs. Most of the studies reported in the literature focused on controlling their magnetic properties, morphology (shape) and size. Additionally, emerging new applications using functionalized $Fe_3O_4$ nanostructures are also highlighted, with emphasis on applications in water treatment, biosensing, ferrofluids, magnetic separation techniques, biomedical technologies as theranostic agents (magnetic resonance imaging [MRI], contrast agents, and drug delivery systems, and magnetic hyperthermia treatment), energy storage, and conversion devices [7–9]. There are many ways to prepare $Fe_3O_4$, such as reduction of hematite by gaseous reductants, energy grinding or preparation in the presence of two aqueous solutions of a ferrous, and ferric salt, using different methods such as co-precipitation, micro-emulsion, and hydrothermal synthesis [10–12].

Carbon is the most widely used material as an absorbent for water purification. Activated carbon, with its large specific surface area and low cost, is the most affordable absorption media in various applications. One of the new additions to the carbon family is graphene, formed by a monolayer of carbon atoms and tightly linked in a hexagonal honeycomb lattice. Layers of graphene stacked on top of each other form graphite, the separate layers of graphene in graphite are held together by van der Waals forces. Graphene has interesting physicochemical properties, light and resistant material, is a good conductor of heat and electricity, is chemically reactive on both sides by grafting of molecules can take the oxide form (graphene oxide GO) and the reduced form (reduced graphene oxide rGO). These different forms of graphene are among the most promising materials. Furthermore, it can be integrated into a large number of applications and used to help improve the performance and efficiency of current materials and substances. It is also used in conjunction with other functional materials to create composites for an even wider range of applications [13–14].

The interplay between the electrical and magnetic properties of $Fe_3O_4$/rGO composites positions them as potential candidates for advancements in various industrial, environmental, and biomedical applications. Composites of reduced graphene oxide and iron oxide have been prepared for different applications and in various methods of preparation. Some applications are gathered in the following **Table 1**.

$Fe_3O_4$/rGO nanocomposites were applied in this study for two interesting applications. Firstly, the elimination of rare earth at low concentration of concentrated phosphoric acid solution, with rapid separation of the metal-loaded magnetic adsorbent from the solution can be achieved using an external magnetic field, and the composite can be regenerated. Secondly to energy storage, their unique properties make them strong contenders for supercapacitor electrodes and lithium-ion battery anodes. These composites offer a captivating blend of high specific capacitance, meaning they can store a large amount of energy in a small space, exceptional energy density, allowing for long-lasting charges, and remarkable cycling stability, ensuring consistent performance over many charge and discharge cycles [28, 29]. Finally, the marriage of $Fe_3O_4$ with rGO increase the surface area to increase the electrochemical reaction sites the graphene sheet prevents the agglomeration of $Fe_3O_4$ nanoparticles resulting in high stability.

In this chapter, we describe the synthesis of $Fe_3O_4$/rGO nanocomposites. A rapid, facile approach and less energy-intensive were developed for synthesis. Structural and morphological properties are determined using various analytical techniques, such as

| Synthesis methods | Applications | Ref. |
|---|---|---|
| Co-precipitation method | – Selective Probes for Pesticide detection by magnetic separation<br>– Targeted anticancer drug delivery | [15–18] |
| Staudenmaier method | Removal of Cr(VI) from aqueous solution | [19] |
| co-precipitation and hydrothermal methods | Photocatalyst for the degradation of different textile dyes | [20] |
| hydrothermal | Magnetic separation | [21] |
| Ultrasonication and chemical reduction | Anode materials for lithium ion storage | [22] |
| spray pyrolysis process | Enzyme immobilization | [23] |
| modified Hummer's process | Supercapacitors | [24] |
| Click Chemistry Reaction | Removing organic dyes from water | [25] |
| planetary ball milling technique | coating material for material surface | [26] |
| gas /liquid interface reaction | Lithium-Ion Batteries | [22] |
| Ultrasonication and chemical reduction | Electrochemical detection of acetaminophen (paracetamol) | [27] |

**Table 1.**
*Summary on the synthesis methods and a few apps of $Fe_3O_4$/reduced graphene oxide composite.*

X-ray diffraction (XRD), scanning electron microscopy (SEM), Fourier-transformed infrared spectroscopy (FT-IR), UV–visible spectroscopy, Laser Diffraction Analysis and Raman spectroscopy. The magnetic properties are studied by electron spin resonance (ESR). This is a good method for studying the magnetic properties of various types of materials. In particular, makes it possible to determine the magnetic state of free or integrated nanomagnets into different matrices. The method is based on the division of electron energy levels by the Zeeman Effect which, is due to the interaction of an external magnetic field with magnetic moments of unpaired electrons. This method is quick and requires very little product, around 20 mg [30]. The ESR spectrum is mainly composed of two absorptions. Low field microwave absorption (LFMA) appears for field values lower than 1500 G, it is flat for paramagnetic and monodomain ferromagnetic compounds and appears for multidomains ferromagnetic compounds. The high field absorption called electronic magnetic resonance (EMR), which can be characterized by:

The shape—symmetrical for paramagnetic compounds, asymmetrical for ferromagnetic compounds, flat curve for diamagnetic compounds.

The resonance field ($H_{res}$)—around 3520 G for paramagnetic compounds, lower than this value for ferromagnetic compounds, and flat response for diamagnetic compounds.

The linewidth ($\Delta H_{pp}$)—used as indices of changes in interactions and magnetic states [30].

Finally, experiments were performed to evaluate the effectiveness of this material in two distinct applications: efficient extraction of various ions in concentrated phosphoric acid, and its use as a high-performance supercapacitor, demonstrating its exceptional capacity and capabilities for energy storage. Metal oxide/reduced graphene oxide (rGO) composites are attracting significant interest for supercapacitor applications due to their potential advantages over conventional materials. These composites offer a combination of high specific capacitance from metal oxides and excellent electrical conductivity and mechanical stability from rGO, leading to improved rate capability, long cycle life, and overall stability [31, 32]. $Fe_3O_4$@rGO composites represent a promising class of supercapacitor electrode materials. They combine the high theoretical capacitance of $Fe_3O_4$ nanoparticles with the favorable electrical properties of rGO sheets [33].

By integrating these results, we aim to provide valuable insights into the promising role of $Fe_3O_4$/RGO nanocomposites in advanced energy and environmental technologies.

## 2. Synthesis and characterization of Fe3O4@rGO nanocomposite

### 2.1 Synthesis procedure of Fe3O4@rGO nanocomposite

The synthesis procedure of $Fe_3O_4$@rGO in three steps. First, the synthesis of $Fe_3O_4$ nanoparticles from the iron sand with the co-precipitation method. Secondly, the synthesis of nanocomposite $Fe_3O_4$@GO by using an autoclave heated to 160°C containing a mixing of $Fe_3O_4$ and D-(+)-glucose. Third, $Fe_3O_4$@rGO was formed by the reduction of GO.

#### *2.1.1 Preparation of Fe3O4 nanoparticles*

The synthesis of magnetic iron oxide ($Fe_3O_4$) nanoparticles was meticulously carried out using a co-precipitation technique, which entailed the controlled reaction of iron(III) ($Fe^{3+}$) and iron(II) ($Fe^{2+}$) ions with a base. Our experimental procedure was divided into three sequential stages for optimal nanoparticle formation.

In the first step, we prepared two distinct solutions. In one beaker, 100 milliliters of a Fe(III) solution was prepared by dissolving hexahydrated ferric chloride ($FeCl_3.6H_2O$ Sigma-Aldrich purity ≥98.0%) in deionized water, ensuring a clear and homogeneous mixture. Similarly, in another beaker, 50 mL of a Fe(II) solution was prepared using heptahydrated ferrous sulfate ($FeSO_4.7H_2O$ Sigma-Aldrich purity ≥99.0%). The two solutions are mixed in a 150 ml Erlenmeyer flask, and heated to a controlled temperature of 70°C. A basic solution of 2 M ammonia was added dropwise to the heated mixture, causing a rapid color change as the iron ions reacted to form the desired $Fe_3O_4$ compound.

As the reaction progressed, the characteristic darkening of the solution indicated the formation of $Fe_3O_4$ nanoparticles (**Figure 1**). The resulting nanoparticles were separated and purified through centrifugation and washing steps. It is crucial to safeguard the Fe(II) solution from oxidation to Fe(III) during the synthesis process. To minimize this issue, it is advisable to employ ferrous sulfate ($FeSO_4$) instead of ferrous nitrate ($Fe(NO_3)_2$) as the Fe(II) source. Ferrous nitrate is more prone to oxidation, which can inadvertently generate unwanted iron oxide byproducts. Moreover, the co-precipitation process using nitrate iron can be more challenging to control, as the rapid precipitation reaction can lead to less predictable outcomes and reduced control over the final product's properties. By opting for ferrous sulfate, you not only ensure a more stable Fe(II) precursor but also facilitate a more controlled and predictable co-precipitation reaction, ultimately contributing to the successful synthesis of high-quality and homogeneous $Fe_3O_4$ nanoparticles.

In an alkaline environment, characteristic of nanoparticle synthesis, the surfaces of synthesized magnetic $Fe_3O_4$ nanoparticles display a negative charge, with an isoelectric point of approximately 6.5 [34]. This negative surface charge facilitates the diffusion of cations into the hydration double layer, which in turn leads to adsorption. The adsorption of cations onto the particles' surface increases the probability of aggregation and particle size growth. To counteract this adsorption, ammonia is used due to its slower ion diffusion compared to sodium ($Na^+$) ions, primarily due to their distinct size difference.

It is worth noting that attaining nanometric-scale magnetic particles in highly concentrated and strong base conditions is a demanding task. To recover the $Fe_3O_4$ nanoparticles, a centrifugation step is carried out at 4000 rotations per minute (rpm) to eliminate excess ammonia by washing the powder with ultrapure water. Subsequently, the $Fe_3O_4$ nanoparticles are dried at a temperature of 60°C for 4 hours. It is essential not to prolong the drying process to minimize the risk of excessive aggregation among the prepared particles.

**Figure 1.**
*Formation of black precipitate of $Fe_3O_4$ nanoparticles.*

### *2.1.2 Synthesis of Fe3O4@GO nanocomposite*

A two-step process was employed for the synthesis of $Fe_3O_4$@GO nanocomposite particles. In the first step, 0.2 g of $Fe_3O_4$ nanoparticles were dispersed in ultrapure water (250 mL) within each of four separate beakers. D-(+)-glucose powder (BioReagent, ≥99.5%, Sigma-Aldrich) was then added to each beaker in varying quantities (2, 4, 6, and 8 g). To ensure homogeneous dispersion of the $Fe_3O_4$ particles, the mixtures were subjected to ultrasonication for 20 minutes. It is crucial to note that an extended sonication period can induce aggregation of the particles. Additionally, heating the dispersions during sonication is not recommended to avoid potential physical and chemical alterations to the glucose molecules. Following ultrasonication, the mixtures were transferred to an autoclave and heated at 160°C for 2 hours. **Table 2** summarizes the employed glucose concentrations and the corresponding designations assigned to the synthesized materials.

This method showed its advantages in cost-effectiveness, simplicity, and reproducibility. This process offers several benefits compared to more complex techniques. It requires minimal specialized equipment, consumes less energy compared to high-temperature methods, and can be easily replicated in different laboratories to yield consistent results.

### *2.1.3 Modification of the Fe3O4@GO surface via (trimethylsilyl) ethylenediamine-triacetic acid (TMSEDTA) functionalization*

The functionalization of $Fe_3O_4$@GO core-shell particles with (trimethylsilyl) ethylenediamine-triacetic acid (TMSEDTA) was achieved through a modified method. In the initial step, 2 mg of $Fe_3O_4$@GO powder was precisely weighed and subsequently dispersed in a pre-mixed solvent consisting of 8 mL of ethanol and 2 mL of deionized (DI) water. The resulting suspension was subjected to ultrasonication for a duration of 60 minutes to promote a uniform and homogeneous distribution of the particles within the solvent mixture. Following sonication, 100 μL of TMSEDTA was carefully introduced dropwise into the suspension. To facilitate the chelation reaction between the surface hydroxyl groups on $Fe_3O_4$@GO and the functional moieties of TMSEDTA, a few drops of glacial acetic acid (AcOH) were added as a catalyst, lowering the reaction medium's pH. The entire mixture was then maintained under continuous agitation for a period of 24 hours to ensure complete functionalization of the $Fe_3O_4$@GO particles.

After the reaction completion, a thorough purification process was undertaken to remove any unbound TMSEDTA molecules. The functionalized $Fe_3O_4$@GO powder was repeatedly rinsed with multiple cycles of ultrapure water using centrifugation or a similar technique to separate the particles from the washing solution. This multistep washing process effectively removes any excess or weakly bound TMSEDTA molecules, ensuring a well-defined and controlled functionalized product.

| Materials | Quantity water (mL) | Quantity of $Fe_3O_4$ (g) | Quantity of glucose (g) |
|---|---|---|---|
| $M_1$ | 250 | 0.2 | 2 |
| $M_2$ | 250 | 0.2 | 4 |

**Table 2.**
*Quantities of reagents used for each material.*

## 2.2 Characterization of Fe3O4@rGO nanocomposite

### *2.2.1 X-ray diffraction analysis*

X-ray diffraction (XRD) is a versatile nondestructive analytical technique used to analyze physical properties such as phase composition, crystal structure, and orientation of powder, solid, and liquid samples. Materials may be single-phase or multiphase mixtures and may contain both crystalline and amorphous components. The comparison of the X-ray diffraction pattern obtained from an unknown sample to the patterns of reference databases allows the identification of the phases.

**Figure 2** presents the XRD patterns of $Fe_3O_4$ and $Fe_3O_4$@rGO composites ($M_1$ and $M_2$). The diffraction peaks observed at 2θ values of 30.0°, 35.0°, 43.1°, 57.0°, and 63.0° correspond to the (220), (311), (400), (511), and (440) crystal planes of magnetite (JCPDS No. 88-0315), confirming the presence of $Fe_3O_4$.

The XRD pattern of $Fe_3O_4$@rGO ($M_1$) exhibits similar peaks to $Fe_3O_4$, although with a noticeable decrease in intensity. This reduction can be attributed to the presence of rGO layers within the composite, potentially obscuring or attenuating the X-ray diffraction signal from $Fe_3O_4$ nanoparticles.

The XRD pattern of $Fe_3O_4$@rGO ($M_2$) displays a broad peak between 12° and 30° (2θ). This broad feature is indicative of diffraction arising from the limited crystallinity and turbostratic nature of the rGO sheets [35, 36]. Additionally, the decrease in the intensity of characteristic $Fe_3O_4$ peaks observed in both M1 and M2 samples aligns

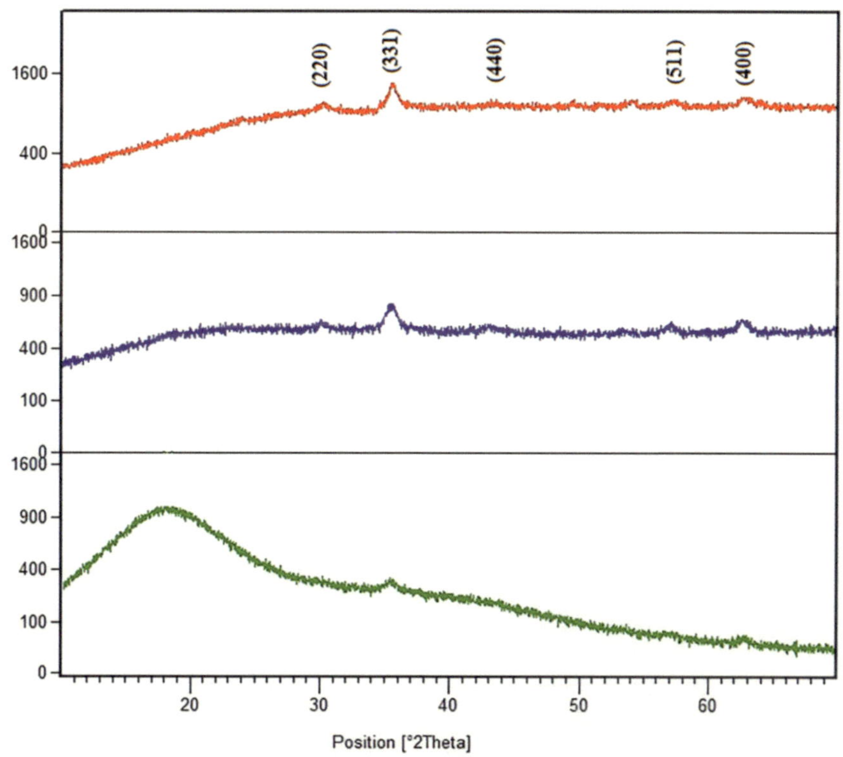


**Figure 2.**
*X-ray diffraction of $Fe_3O_4$ NPs (Red) $Fe_3O_4$@rGO $M_1$ (Blue) $Fe_3O_4$@rGO $M_2$ (Green).*

with the expected decrease in the mass fraction of iron oxide due to the incorporation of rGO.

$M_1$ and $M_2$ represent composite nanomaterials containing two distinct phases: a continuous rGO matrix and dispersed $Fe_3O_4$ nanoparticles. The relative percentage of $Fe_3O_4$ within these composites can be controlled by the preparation method, with higher glucose concentrations during synthesis typically leading to lower $Fe_3O_4$ content.

The Debye-Scherrer equation was employed to estimate the average crystallite size within $Fe_3O_4$ and $Fe_3O_4$@rGO composites ($M_1$ and $M_2$) based on the broadening observed in their respective XRD peaks (**Figure 2**). The analysis revealed an increase in the average crystallite size from 56 nm for $Fe_3O_4$ to a value of approximately 86 nm for both $Fe_3O_4$@rGO ($M_1$) and $Fe_3O_4$@rGO ($M_2$). Further investigations are required to elucidate the definitive cause of this observed crystallite size increase. The augmentation in crystallite size after coating the $Fe_3O_4$ by rGO is attributed to the coalescence of the particles by solid-state diffusion during the transformation process of glucose into reduced graphene oxide at 150°C [30].

### *2.2.2 Laser diffraction analysis*

To further assess the particle size distribution of the $Fe_3O_4$ nanoparticles, laser diffraction granulometry was employed (**Figure 3**). The resulting particle size distribution, presented as a function of particle number, reveals the presence of two distinct populations. The dominant population exhibits an average size of approximately 82 nm. However, a significantly smaller population with a much larger average size of around 500 nm is also observed. This substantial difference in size and quantity suggests that the second population likely consists of aggregated $Fe_3O_4$ nanoparticles.

### *2.2.3 Micromorphology and elemental analysis SEM-EDX*

**Figure 4** presents high-resolution scanning electron microscopy (SEM) images and corresponding energy-dispersive X-ray spectroscopy (EDX) spectra of the $Fe_3O_4$@rGO composites ($M_1$ and $M_2$). The SEM images reveal a well-defined layered morphology for both $M_1$ and $M_2$. This observation suggests the successful formation

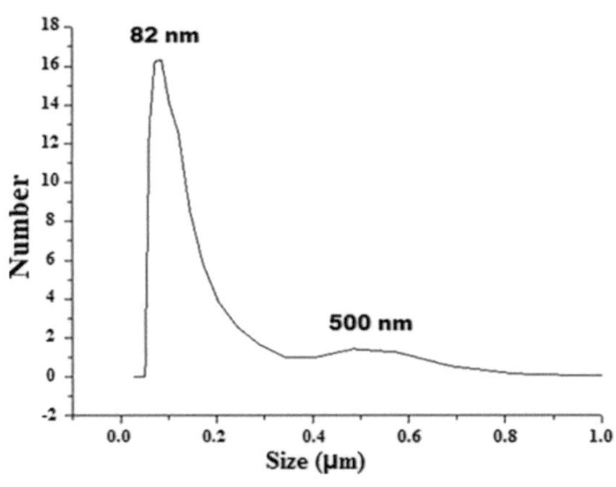


**Figure 3.**
*Particle size distribution of $Fe_3O_4$.*

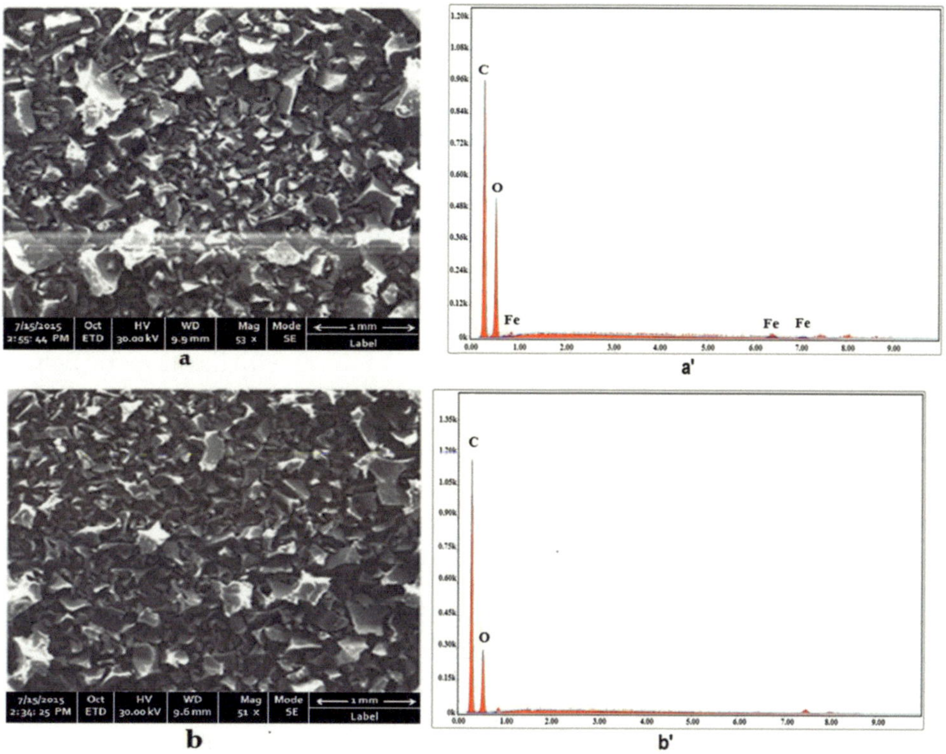


**Figure 4.**
*SEM image of $M_1$ (a) and $M_2$ (b), EDX spectrum of $M_1$ (a') and $M_2$ (b').*

of a multi-layered structure composed of reduced graphene oxide (rGO) sheets. Notably, individual $Fe_3O_4$ nanoparticles are not readily identifiable within the SEM images. This could be attributed to two potential factors:

Encapsulation: The $Fe_3O_4$ nanoparticles might be encapsulated within the folded or crumpled layers of the rGO sheets, hindering their direct visualization through SEM. Particle size and resolution limits: The size of the $Fe_3O_4$ nanoparticles employed in the synthesis might be below the resolution limit of the SEM instrument, making them difficult to distinguish from the surrounding rGO matrix.

The EDX analysis provides further insights into the elemental composition of the composites. The EDX spectrum of $Fe_3O_4$@rGO ($M_1$) exhibits a weak iron (Fe) signal (**Figure 4a'**). This suggests the presence of some $Fe_3O_4$ within the composite, potentially encapsulated by the rGO layers. In contrast, the EDX spectrum of $Fe_3O_4$@rGO (M2) does not show any detectable iron peak (**Figure 4b'**). This significant difference between M1 and M2 can be attributed to the following: Thicker rGOshell: The synthesis process of M2 leads to the formation of a thicker rGO shell than that of M1. This thicker shell completely masks the $Fe_3O_4$ nanoparticles. Lower $Fe_3O_4$ content: The synthesis parameters of $M_2$ result in a composite with a lower relative overall $Fe_3O_4$ content than that of $M_1$.

The results reveal a direct correlation between the carbon content observed in the synthesized materials and the amount of glucose utilized during the fabrication process. The higher glucose concentration employed for sample $M_2$ likely contributed to the formation of a greater number of graphene oxide layers surrounding the $Fe_3O_4$ nanoparticles.

### *2.2.4 Magnetic properties*

Magnetite exhibits ferromagnetic (FM) behavior. It can exhibit the strongest magnetism among other iron oxide phases. At the nanometric scale, $Fe_3O_4$ nanoparticles have new magnetic properties that are influenced by their size, their morphology, and their crystallinity. These parameters are strongly affected by the synthesis methods. We focus on the analysis of the magnetic characteristics of $Fe_3O_4$ NPs at different stages of synthesis due to the particular interest in their use in various technologies and applications. For crystallite, size below 25 nm $Fe_3O_4$ is in a superparamagnetic state (SPM) and between 25 and 80 nm is the single-domain magnetite (SD), and it is in a multidomain magnetic (MD) state for larger sizes [25]. **Figure 5** illustrates the relationship of the $Fe_3O_4$ NP size (diameter) with its magnetic behavior and coercivity ($H_c$).

**Figure 6** displays the electron spin resonance (ESR) spectra obtained at ambient temperature for the samples $Fe_3O_4$, $Fe_3O_4$@rGO ($M_1$) and $Fe_3O_4$@rGO ($M_2$).

The spectrum of the compound $Fe_3O_4$ is formed by a single line, the resonance field Hr 3356 Gauss (**Table 3**), a value lower than that observed for a free electron (3520 Gauss), suggesting the existence of internal fields resulting from ferrimagnetic ordering. The observed sample displayed a significant line width ($\Delta H_{pp}$ = 1133 G), which may be explained by the presence of a random distribution of the anisotropy axis in single-domain nanoparticles ferromagnetic. In addition, the absence at low frequency of the absorption magnetic absorption (LFMA) signal clearly shows that $Fe_3O_4$ is in a magnetic single domain (MD) state [7]. The magnetic state agrees well with the size of the determined crystallites which is of the order of 56 nm. ESR spectra of $M_1$ and $M_2$ have the same line shape. The crystallite size of iron oxide nanoparticles in M1 and M2 are respectively equal to 86 and 88 nm 86 nm, values above the size threshold required for the transition from a single domain to a multidomain magnetic state (MD). The ESR spectra of $M_1$ and $M_2$ are in agreement with these results. These two spectra have the same line shape and the spectroscopic parameters, namely $H_{res}$ $\Delta H_{pp}$ are identical, with the appearance of absorption at low fields. A shift toward weak fields of $H_{res}$ = 2950 Gauss is in good agreement with the increase in crystallite size. The increase in crystallite size is attributed to the coalescence of particles by diffusion during the process of transformation of glucose into graphene oxide at 150°C [37]. The appearance of broad absorption toward weak fields (<1500 G) confirms the

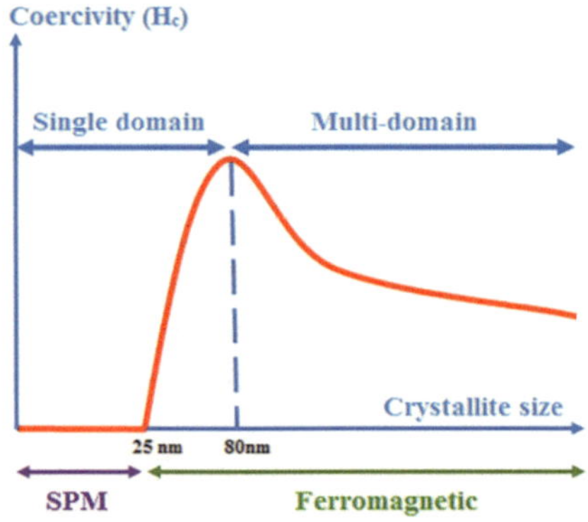


**Figure 5.**
*Relationship between the size of nanomagnet crystallites and their magnetic behavior and coercivity, critical sizes of magnetic change of $Fe_3O_4$ are shown.*

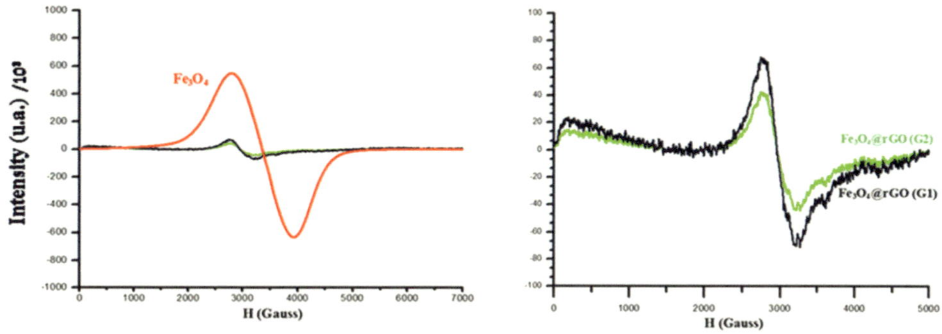


**Figure 6.**
*ESR spectra of $Fe_3O_4$ Red, $Fe_3O_4$@rGO Black ($M_1$) and $Fe_3O_4$@rGO Green ($M_2$).*

| Samples | $H_{res}$ (Gauss) | $\Delta H$ (Gauss) | $\Delta I_{pp}/10^3$ | D (nm) |
|---|---|---|---|---|
| $Fe_3O_4$ | 3358 | 1133 | 1182.10 | 56 |
| $Fe_3O_4$@rGO ($M_1$) | 2950 | 447 | 413.10 | 86 |
| $Fe_3O_4$@rGO ($M_2$) | 2950 | 471 | 171.03 | 88 |

**Table 3.**
*ESR parameters and crystallite sizes of $Fe_3O_4$ in the different samples.*

multidomain state. This nonresonant absorption is attributed to the displacement of the walls separating the magnetic domains [38].

### *2.2.5 Raman spectroscopy*

Raman spectroscopy was used to validate the successful encapsulation of $Fe_3O_4$ nanoparticles by reduced graphene oxide (rGO). As shown in **Figure 7**, the Raman spectra of $Fe_3O_4$ nanoparticles and $Fe_3O_4$@GO composites show significant differences. The $Fe_3O_4$@GO (M1 and M2) spectrum clearly shows the D, G bands at 1351 and 1606 $cm^{-1}$, respectively, which are the characteristic bands of graphene. The intensity ratio of the D and G bands, ID/IG provides insight into the reduction process by elimination of the oxygen functional groups that create imperfections within the carbon. The Raman spectrum of graphene oxide (GO) gives a D band with an intensity comparable to the G band [39]. The ID/IG ratio for reduced graphene oxide is generally less than one which is the case for the two compounds $Fe_3O_4$@GO ($M_1$ and $M_2$), which confirms the formation of rGO [40]. The G band shift (observed in $M_2$ compared to $M_1$) toward a higher wavenumber (1370 $cm^{-1}$) suggests a degree of graphitization within the composites [41].

### *2.2.6 XPS investigation of silanization process*

X-ray photoelectron spectroscopy (XPS) provides elemental composition and chemical state information of the material's topmost layers (typically 0–10 nm). This analysis was performed to investigate the grafting of silane molecules (TMSEDTA) onto $Fe_3O_4$@GO nanocomposite surfaces. **Figure 8** depicts the spectra of $Fe_3O_4$@GO before and after silanization with TMSEDTA. A significant

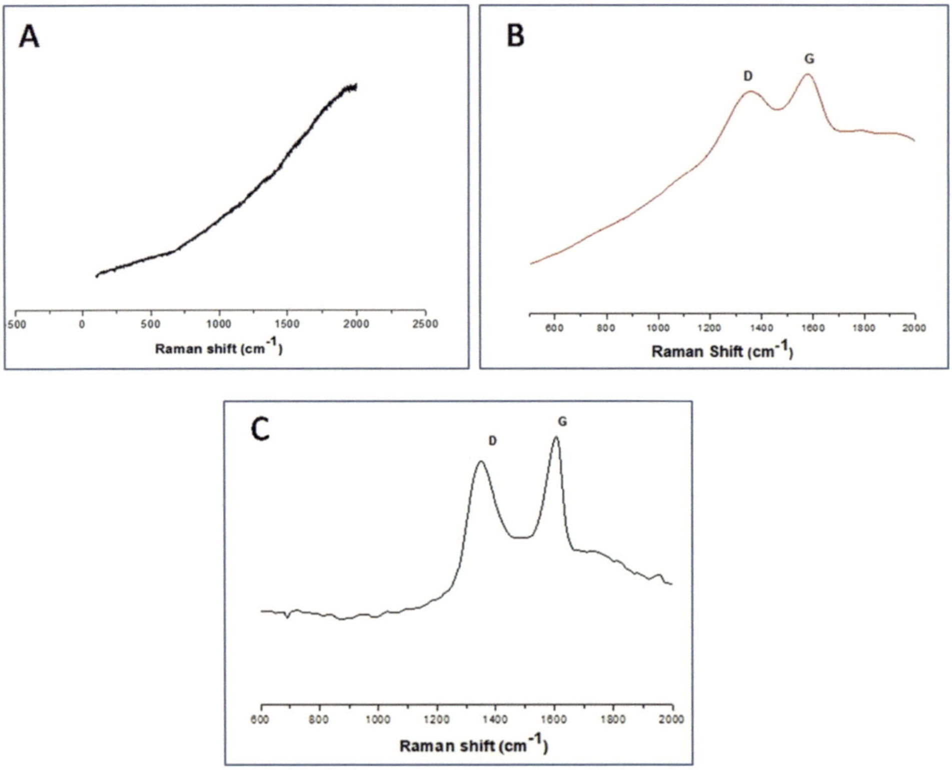


**Figure 7.**
*Raman spectra of $Fe_3O_4$ (A), Raman spectra of $Fe_3O_4$@rGO ($M_1$), Raman spectra of $Fe_3O_4$@rGO ($M_2$).*

alteration in the surface chemistry is observed upon silanization. The most prominent feature in the $Fe_3O_4$@GO@TMSEDTA spectrum (**Figure 8A**) is the emergence of new peaks corresponding to the characteristic binding energies of Na (1 s), C (1 s), N (1 s), and Si (2p) elements. These elements are absent in the $Fe_3O_4$@GO spectrum (**Figure 8B**), confirming the presence of grafted TMSEDTA molecules. **Table 4** summarizes the quantitative analysis of the surface elemental composition obtained from XPS data. A notable increase in the carbon (C) content is observed after silanization with TMSEDTA. The initial carbon content in $Fe_3O_4$@GO originates from the presence of graphene oxide. The observed elevation in the surface carbon proportion between $Fe_3O_4$@GO and $Fe_3O_4$@GO@TMSEDTA, along with the presence of newly emerged peaks for Na, N, and Si, strongly suggests the successful grafting of TMSEDTA onto the graphene oxide surface. This functionalization process introduces silane groups (Si-O-), enhancing the compatibility of the material for subsequent applications. Furthermore, the disappearance of the Fe $2p^3$ peak in the $Fe_3O_4$@GO@TMSEDTA spectrum (**Figure 8B**) can be attributed to the formation of a thick TMSEDTA layer on the $Fe_3O_4$@GO core-shell structure. This observation implies the formation of a multilayer TMSEDTA coating on the surface, effectively encapsulating the $Fe_3O_4$ nanoparticles. The thickness of this layer can be further investigated using higher-resolution XPS analysis techniques or complementary characterization methods like transmission electron microscopy (TEM).

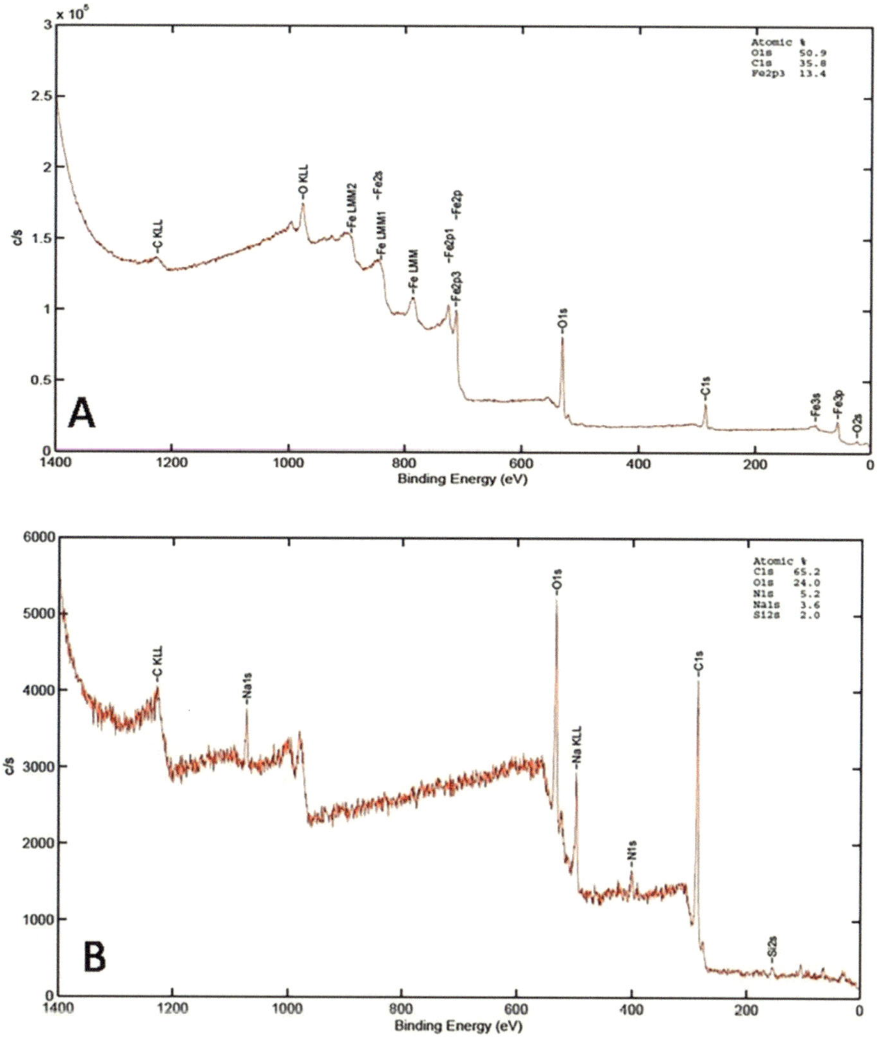


**Figure 8.**
*XPS analysis of $Fe_3O_4$@GO (A), XPS analysis of $Fe_3O_4$@GO@TMSEDTA (B).*

| | C1s | O1s | $Fe2p^3$ | N1s | Si2s | Na1s |
|---|---|---|---|---|---|---|
| $Fe_3O_4$@GO | 35.8 | 50.9 | 13.4 | 0 | 0 | 0 |
| $Fe_3O_4$@GO@TMSEDTA | 65.2 | 24.0 | 0 | 5.2 | 3.6 | 2.0 |

**Table 4.**
*Quantitative analysis of surface composition.*

## 3. Extraction of rare earth elements from phosphoric acid using Fe3O4@ GO@TMSEDTA

This section investigates the efficiency of $Fe_3O_4$@GO@TMSEDTA for extracting rare earth elements (REEs) from phosphoric acid solutions at varying pH levels (diluted, pH 4 and 2.5; concentrated). The study aims to assess the dependence of extraction efficiency on solution acidity and individual REE electronic configurations.

### 3.1 Experimental details and results

Experiments involved mixing 2 mg of $Fe_3O_4$@GO@TMSEDTA with 10 mL of $H_3PO_4$ solutions at different pH levels, spiked with a defined amount of a Gd/Eu mixture (total REE concentration 35 mg/L). After vigorous stirring for 12 hours and magnetic separation of $Fe_3O_4$@GO@TMSEDTA, the remaining solution was analyzed by ICP-OES to determine the residual REE concentration. **Tables 5–7** summarize the initial and final REE concentrations for each pH level.

The results indicate high overall efficiency of $Fe_3O_4$@GO@TMSEDTA for REE extraction across all tested conditions. A slight decrease in extraction efficiency was observed for some REEs at lower pH (likely due to decreased negative surface charge density on $Fe_3O_4$@GO@TMSEDTA below its point of zero charge). Notably, complete extraction was achieved for Praseodymium and Terbium at all pH levels, while Neodymium, Samarium, and Cerium displayed efficiencies exceeding 90%. The remaining REEs exhibited varying efficiencies that did not exceed 85%.

While all extraction percentages exceeded 70%, they are not solely indicative of process efficiency or selectivity due to differences in initial REE concentrations. Analysis of the amount of each REE retained per gram of $Fe_3O_4$@GO@TMSEDTA provides a more relevant comparison. For instance, the higher retention quantity of Cerium compared to Samarium, despite similar extraction percentages, can be attributed to the electronic configuration of each element influencing vacant site availability for complexation.

Similarly, Europium's higher retention quantity compared to Erbium, despite lower extraction percentage, can be explained by its favorable electron configuration for stable complex formation. This competitive chelation process, influenced by relative chelation rates, determines which REEs are preferentially extracted until $Fe_3O_4$@GO@TMSEDTA reaches its maximum capacity. Interestingly, gadolinium displays the lowest extraction percentages but the highest overall retention quantity.

| Rare earths | Initial C. ($mg\,L^{-1}$) | C. after extraction ($mg\,L^{-1}$) | Quantity retained ($mg\,g^{-1}$) | Extraction % |
|---|---|---|---|---|
| Samarium Sm | 8.600 | 0.450 | 40.75 | 94.76 |
| Europium Eu | 39.700 | 11.798 | 137.50 | 70.28 |
| Gadolinium Gd | 47.500 | 15.550 | 159.75 | 67.26 |
| Cerium Ce | 13.250 | 1.650 | 58.00 | 87.54 |
| Praseodymium Pr | 1.300 | 0 | 6.50 | 100 |
| NeodymiumNd | 16.875 | 1.700 | 75.87 | 89.92 |
| Erbium Er | 4.711 | 1.125 | 17.93 | 76.12 |
| Terbium Tb | 1.120 | 0 | 5.60 | 100 |
| Thulium Th | 3.125 | 0.920 | 11.25 | 72 |

**Table 5.**
*Concentrations and quantities of rare earths after extraction of concentrated phosphoric acid.*

| Rare earths | Initial C. (mg L$^{-1}$) | C. after extraction (mg L$^{-1}$) | Quantity retained (mg g$^{-1}$) | Extractin % |
|---|---|---|---|---|
| Samarium Sm | 8.600 | 0.250 | 41.75 | 97.09 |
| Europium Eu | 39.700 | 7.000 | 163.50 | 82.36 |
| Gadolinium Gd | 47.500 | 9.550 | 189.75 | 79.89 |
| Cerium Ce | 13.250 | 0.350 | 64.50 | 97.35 |
| Praseodymium Pr | 1.300 | 0.001 | 6.49 | 99.92 |
| Neodymium Nd | 16.875 | 1.210 | 78.32 | 92.82 |
| Erbium Er | 4.711 | 0.725 | 19.95 | 84.61 |
| Terbium Tb | 1.120 | 0 | 5.60 | 100 |
| Thulium Tm | 3.125 | 0.585 | 12.70 | 81.28 |

**Table 6.**
*Concentrations and quantities of rare earths before and after phosphoric acid extraction pH = 4.*

| Rare earths | Initial C. (mg L$^{-1}$) | C. after extraction (mg L$^{-1}$) | Quantity retained (mg g$^{-1}$) | Extraction % |
|---|---|---|---|---|
| Samarium Sm | 8.600 | 0.325 | 41.35 | 96.22 |
| Europium Eu | 39.700 | 8.500 | 156.00 | 78.85 |
| Gadolinium Gd | 47.500 | 11.350 | 180.07 | 76.10 |
| Cerium Ce | 13.250 | 0.500 | 63.75 | 96.21 |
| PraseodymiumPr | 1.300 | 0.000 | 6.50 | 100 |
| NeodymiumNd | 16.875 | 1.190 | 78.42 | 92.94 |
| Erbium Er | 4.711 | 0.889 | 19.11 | 81.12 |
| Terbium Tb | 1.120 | 0 | 5.60 | 100 |
| Thulium Tm | 3.125 | 0.688 | 12.18 | 77.98 |

**Table 7.**
*Concentrations and quantities of rare earths before and after phosphoric acid extraction pH = 2.5.*

## 3.2 Extraction from concentrated phosphoric acid

**Table 5** demonstrates a significant reduction in final REE concentrations for concentrated phosphoric acid compared to samples at pH 4 and 2.5 (**Tables 6** and 7). Despite this decrease, high extraction efficiencies were still achieved for some REEs (e.g., Samarium, Neodymium, and Cerium) considering their initial concentrations. This decrease is expected due to the operating pH being likely below the point of zero charge of $Fe_3O_4$@GO@TMSEDTA. Notably, the material effectively removed all detectable quantities of Praseodymium and Terbium present in the solution, likely due to their very low initial concentrations.

The separation of Erbium and Thulium was achieved with lower efficiencies, while Gadolinium and Europium displayed moderate efficiencies. However, considering their initial concentrations, $Fe_3O_4$@GO@TMSEDTA demonstrates a strong capacity for removing significant quantities of both Europium and Gadolinium, alongside other REEs present in concentrated phosphoric acid solutions.

### 3.3 Proposed mechanism of elimination

The elimination process might proceed in two stages. The first stage likely involves the chelation of RE ions by the carboxylic acid functionalities of the grafted silane molecules via Van der Waals forces or electrostatic interactions. The second stage involves the magnetic separation of the $Fe_3O_4$@GO@TMSEDTA nanocomposites from the phosphoric acid solution using a magnet.

$Fe_3O_4$@GO@TMSEDTA demonstrates the promising potential for extracting REEs from phosphoric acid solutions across a range of pH levels. The observed variations in extraction efficiency and retention quantities highlight the influence of both solution pH and individual REE electronic configurations on the interaction with the functionalized material. Notably, the material exhibited high efficiency for some REEs even at a pH likely below its point of zero charge. This suggests the effectiveness of $Fe_3O_4$@GO@TMSEDTA for REE extraction under various practical conditions. Further studies could optimize the process by exploring the effect of other parameters like contact time, temperature, and the initial REE concentration ratio to achieve even higher extraction efficiency and selectivity for specific REEs.

## 4. Impact of Fe3O4 nanoparticle size on the electrochemical performance of Fe3O4@rGO supercapacitor electrodes

This study investigates the influence of $Fe_3O_4$ nanoparticle size on the electrochemical performance of $Fe_3O_4$@rGO composites (denoted as $G_1$ and $G_2$) as supercapacitor electrodes.

### 4.1 Electrode fabrication

A two-step process was employed to fabricate the electrodes. Two milligrams of either $M_1$ or $M_2$ powder was uniformly distributed onto a 0.5 $cm^2$ nickel foam current collector. The powder-nickel foam assembly was then pressed at 2 bar using a manual press machine. The exposed working electrode area for each sample was approximately 0.5 $cm^2$.

### 4.2 Electrochemical characterization

Electrochemical measurements were performed using a conventional three-electrode system in a 0.1 M LiOH electrolyte solution at room temperature. A platinum sheet with a surface area of 0.5 $cm^2$ served as the counter electrode, and an Ag/AgCl electrode served as the reference electrode. Cyclic voltammetry (CV) experiments were conducted within a potential range of −0.1 to +0.4 V for $M_1$ and −0.1 to +0.55 V for $M_2$ at various scan rates (50, 100, 150, and 200 mV/s).

### 4.3 Capacitance calculation

The specific capacitance ($C_p$) of the electrodes was calculated using the following equation based on the CV curves: $C_p = (i \times \Delta t) / (\Delta V \times m)$.

$C_p$: Specific capacitance (F)

I: Current (A)

$\Delta t$: Scan time (s)

ΔV: Potential window (V)

m: Mass of active material (g)

The CV curves presented in **Figure 9** reveal distinct electrochemical behaviors for $G_1$ and $G_2$. The first containing a higher $Fe_3O_4$ content with a slightly smaller crystallite size, exhibits a narrower potential window (1.4 V) from −1 to 0.4 V but a higher specific capacitance of 428.57 F/g. Conversely, $M_2$ exhibits a wider potential window (1.55 V) ranging from −1 to 0.55 V but a lower specific capacitance of 235.48 F/g. This difference can be attributed to the increased rGO content and decreased $Fe_3O_4$ content in $M_2$.

These observations are consistent with theoretical predictions regarding the influence of $Fe_3O_4$ nanoparticle size on capacitance [42, 43]. Smaller nanoparticles provide a larger active surface area and more reaction sites for electrochemical processes, potentially leading to higher capacitance. However, smaller dimensions can also limit the electrochemical window due to increased particle aggregation and instability at higher potentials. In contrast, larger nanoparticles offer improved stability and a broader potential window, but at the expense of reduced surface area and fewer active sites, resulting in lower capacitance.

Cycling stability tests were conducted on $M_1$ and $M_2$ over 3000 cycles. M1 demonstrated superior stability with minimal specific capacitance loss (8.5%, from 428.57 to 392.85 F/g). In contrast, M2 exhibited a significant decrease in specific capacitance (over 50%, from 235.48 to 135.71 F/g).

The observed differences in electrochemical performance can be attributed to the influence of $Fe_3O_4$ nanoparticle size on magnetic properties. When nanoparticle size exceeds a certain threshold, a transition from single-domain to multidomain ferromagnetic behavior occurs. The larger $Fe_3O_4$ nanoparticles in $M_2$ may undergo

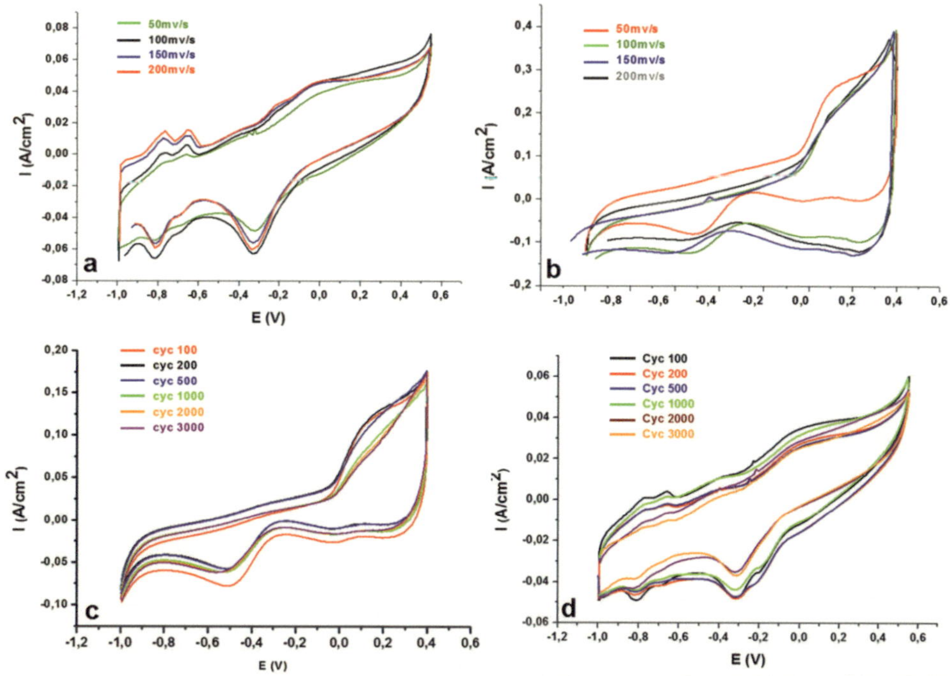


**Figure 9.**
*Cyclic voltammetry measurements of $Fe_3O_4$@rGO ($G_1$) (b,c) and $Fe_3O_4$@rGO($G_2$) (a,d) [16].*

this transition, leading to reduced stability and capacity retention while expanding the potential window. Conversely, the smaller, nanoparticles in $M_1$ contribute to enhanced stability and improved capacity retention.

This study demonstrates the critical role of $Fe_3O_4$ nanoparticle size in $Fe_3O_4$@rGO supercapacitor electrodes. Smaller $Fe_3O_4$ nanoparticles ($M_1$) yielded higher capacitance (428.57 F/g) due to increased surface area but exhibited a narrower potential window. Conversely, larger nanoparticles ($M_2$) offered a wider window but lower capacitance (235.48 F/g). M1 also displayed superior cycling stability, highlighting the importance of optimizing nanoparticle size for balanced performance in supercapacitor applications.

## 5. Conclusion

In conclusion, $Fe_3O_4$@rGO is synthesized by a three-step process involving co-precipitation synthesis of $Fe_3O_4$, the coating of the latter with graphene oxide prepared from glycose and reduction of graphene oxide (rGO). Synthetic materials are studied during the different stages of synthesis by XRD, XRD, and SEM. ESR measurements show a change in the magnetic state of iron oxide during the synthesis process. Electrochemical tests showed that the prepared $Fe_3O_4$@rGO had a high specific capacitance of 428.57 F/g and remarkable stability after 3000 cycles. The results demonstrated the great potential of the $Fe_3O_4$@rGO composite for energy storage applications. TMSEDTA molecules grafting $Fe_3O_4$@rGO gives the nanocomposite TMSEDTA@GO@$Fe_3O_4$ which, having high resistance to corrosion in a strongly acidic environment phosphoric acid, and used for the extraction of rare earth metals in phosphoric acid. These nanocomposites can offer unique advantages for various applications, including environmental and economic purposes by extraction of rare earth from phosphoric acid and for energy storage applications.

## Author details

Mondher Yahya* and Haythem Barrak
Laboratory of Useful Materials Valuation, National Center for Research in Materials Sciences, Soliman, Tunisia

*Address all correspondence to: mondher.y@gmail.com; mondher.yahya@cnrsm.rnrt.tn

## References

[1] Imran M, Chaudhary AA, Ahmed S, Alam MM, Khan A, Zouli N, et al. Iron oxide nanoparticle-based ferro-nanofluids for advanced technological applications. Molecules. 2022;**27**(22):7931. DOI: 10.3390/molecules27227931

[2] Asma G, Hamdi N, Rana S, Obada AS, Aseel AR, Alaa SS, et al. Magnetic iron oxide-based nanozymes: From synthesis to application. Nanoscale Advances. 2024;**24**(6):1611-1642. DOI: 10.1039/D3NA00903C (Review Article)

[3] Lafta SH. Evaluation of hematite nanoparticles weak ferromagnetism. Journal of Superconductivity and Novel Magnetism. 2020;**33**:3765-3772. DOI: 10.1007/s10948-020-05626-8

[4] Nelson N, Ruth P, Carlo M, Jacobo H-M, Miguel MS. First-principles calculations of hematite (α-Fe2O3) by self-consistent DFT+U+V. iScience. 2023;**26**:106033. DOI: 10.1016/j.isci.2023.106033

[5] Mounkachi O, Lamouri R, Hamedoun M, Ez-Zahraouy H, Salmani E, Benyoussef A. Effect of defects disorder on the half-metallicity, magnetic properties, and gap states of $Fe_3O_4$: A first principles study. Journal of Superconductivity and Novel Magnetism. 2017;**30**:3221-3224. DOI: 10.1007/s10948-017-4108-3

[6] Shivani S, Blas PU, Mark A. Density functional theory study of local environment effects on oxygen vacancy properties in magnetite. The Journal of Physical Chemistry C. 2023;**127**(35):17460-17472. DOI: 10.1021/acs.jpcc.3c02581

[7] Minh DN, Hung-Vu T, Shoujun X, Randall L. $Fe_3O_4$ nanoparticles: Structures, synthesis, magnetic properties, surface functionalization, and emerging applications. Applied Sciences. 2021;**11**(23):11301. DOI: 10.3390/app112311301

[8] Chen Y, Ding X, Zhang Y, Natalia A, Sun X, Wang Z, et al. Design and synthesis of magnetic nanoparticles for biomedical diagnostics. Quantitative Imaging in Medicine and Surgery. 2018;**8**(9):957-970. DOI: 10.21037/qims.2018.10.07

[9] Amraee A, Alamzadeh Z, Irajirad R, et al. Theranostic RGD@$Fe_3O_4$-Au/Gd NPs for the targeted radiotherapy and MR imaging of breast cancer. Cancer Nanotechnology. 2023;**14**:61. DOI: 10.1186/s12645-023-00214-6

[10] Jianwen Y, Yuexin H, Yanjun L, Peng G, Wenbo L. Mechanism and kinetics of the reduction of hematite to magnetite with CO–CO2 in a micro-fluidized bed. Minerals. 2017;7(11):209. DOI: 10.3390/min7110209

[11] Muneer M, Abbad BA, Abdelbaki B, Dina E, Abdul WM, Ebrahim M. Synthesis of $Fe_3O_4$ nanoparticles with different shapes through a co precipitation method and their application. JOM. 2022;**74**(9):1-9. DOI: 10.1007/s11837-022-05380-3

[12] Ganesha A, Putri DJ, Yus RD, Andri S. Utilization of co-precipitation method on synthesis of $Fe_3O_4$/PEG with different concentrations of PEG for biosensor applications. Materialia. 2022;**25**:101525. DOI: 10.1016/j.mtla.2022.101525

[13] Kamboj S, Thakur A. Applications of graphene-based composites—A review. Materials Today Proceedings. 2023. In press. DOI: 10.1016/j.matpr.2023.01.041

[14] Zheng W, Zhao X, Fu W. Review of Vertical Graphene and its Applications. ACS Applied Materials & Interfaces. 2021;**13**:9561-9579. DOI: 10.1021/acsami.0c19188

[15] Yukird J, Insin N, Chanajaree R, Rodthongkum N. $Fe_3O_4$ nanoparticle/graphene oxide composites as selective probes and self-matrixes for pesticide detection by electrochemistry and laser desorption/ionization mass spectrometry. ACS Applied Nano Materials. 2023;**6**(13):11912-11924. DOI: 10.1021/acsanm.3c01762

[16] Wang LH, Sui L, Zhao PH, Ma HD, Liu JY, Wei Z, et al. A composite of graphene oxide and iron oxide nanoparticles for targeted drug delivery of temozolomide. Die Pharmazie. 2020;**75**(7):313-317. DOI: 10.1691/ph.2020.9170

[17] Wang J, Fang J, Fang P, Li X, Wu S, Zhang W, et al. Preparation of hollow core/shell $Fe_3O_4$@graphene oxide composites as magnetic targeting drug nanocarriers. Journal of Biomaterials Science. Polymer Edition. 2017;**28**(4):337-349. DOI: 10.1080/09205063.2016.1268463

[18] Kohzadi S, Najmoddin N, Baharifar H, Shabani M. Functionalized SPION immobilized on graphene-oxide: Anticancer and antiviral study. Diamond and Related Materials. 2022;**127**:109149. DOI: 10.1016/j.diamond.2022.109149

[19] Mahvi AH, Balarak D, Bazrafshan E. Remarkable reusability of magnetic $Fe_3O_4$-graphene oxide composite: A highly effective adsorbent for Cr(VI) ions. International Journal of Environmental Analytical Chemistry. 2021;**103**(15):3501-3521. DOI: 10.1080/03067319.2021.1910250

[20] Baptisttella AMS, Araujo C, da Silva MP, do Nascimento GFO, Costa GPD, do Nascimento BF, et al. Magnetic $Fe_3O_4$-graphene oxide nanocomposite—Synthesis and practical application for the heterogeneous photo-fenton degradation of different dyes in water. Separation Science and Technology. 2020;**56**(2):425-438. DOI: 10.1080/01496395.2020.1716011

[21] Zhang F, Xue X, Huang I, Yang H. Adsorption and heterogeneous fenton catalytic performance for magnetic $Fe_3O_4$/reduced graphene oxide aerogel. Journal of Materials Science. 2020;**55**:15695-15708. DOI: 10.1007/s10853-020-05159-4

[22] Bhuvaneswari S, Pratheeksha PM, Anandan S, Rangappa D, Gopalan R, Rao TN. Efficient reduced graphene oxide grafted porous $Fe_3O_4$ composite as a high performance anode material for Li-ion batteries. Physical Chemistry Chemical Physics. 2014;**16**:5284-5294. DOI: 10.1039/C3CP54778G

[23] Patel SKS, Choi SH, Kang YC, Lee J-K. Eco-friendly composite of $Fe_3O_4$-reduced graphene oxide particles for efficient enzyme immobilization. ACS Applied Materials and Interfaces. 2017;**9**(3):2213-2222. DOI: 10.1021/acsami.6b05165

[24] Rosaiah P, Zhu J, Hussain OM, et al. Facile and cost-effective synthesis of flower-like RGO/$Fe_3O_4$ nanocomposites with ultra-long cycling stability for supercapacitors. Ionics. 2019;**25**:655-664. DOI: 10.1007/s11581-018-2669-1

[25] Hao Y, Wang Z, Gou J, Wang Z. Kinetics and thermodynamics of diquat removal from water using magnetic graphene oxide nanocomposite. The Canadian Journal of Chemical Engineering. 2015;**93**(1):1713-1720. DOI: 10.1002/cjce.22278

[26] Dash T, Katika AK, Palei BB, Dhal JP, Rout TK. Preparation of iron

oxide-graphene oxide composites and their characterizations. Materials Today: Proceedings. 2021;**44**(Part 6):4657-4660. DOI: 10.1016/j.matpr.2020.10.934

[27] Shen L, Dong J, Wen B, Wen X, Li J. Facile synthesis of hollow $Fe_3O_4$-rGO nanocomposites for the electrochemical detection of acetaminophen. Nanomaterials. 2023;**13**:707. DOI: 10.3390/nano13040707

[28] Haythem B, Ridha A, Pascale C, Adel M, Gaétan L, Ahmed HH. Highly efficient extraction of rare earth elements and others ions from green phosphoric acid medium using TMSEDTA@ GO@$Fe_3O_4$ core-shell. Separation and Purification Technology. 2019:145-151. DOI: 10.1016/j.seppur.2019.04.016

[29] Haythem B, Mondher Y, Ridha A, Ahmed HH. Preparation and magnetic properties of $Fe_3O_4$@ rGO nanocomposite: Application in aqueous electrochemical supercapacitor. Chemistry Africa. 2024. In press. DOI: 10.1007/s42250-024-00955-w

[30] Mondher Y, Faouzi H, Ahmed HH. Synthesis and ESR study of transition from ferromagnetism to superparamagnetism in La0.8Sr0.2MnO3 nanomanganite. In: Tatyana S, Vladimir B, editors. Handbook of Smart Nanosystems for Biomedicine, Optoelectronics and Catalysis. Russia: Moscow State University; DOI: 10.5772/ intechopen.89951

[31] Saha S, Jana M, Samanta P, Murmu NC, Kim NH, Kuila T, et al. Hydrothermal synthesis of $Fe_3O_4$/ RGO composites and investigation of electrochemical performances for energystorage applications. RSC Advances. 2014;**4**:44777-44785. DOI: 10.1039/C4RA07388F

[32] Liao J, Li Y, Wang Z, Lv L, Chang L. In-situ preparation of $Fe_3O_4$/ graphene nanocomposites and their electrochemical performances for supercapacitor. Materials Chemistry and Physics. 2021;**258**:123995. DOI: 10.1016/j. matchemphys.2020.123995

[33] Maharsi R, Arif AF, Ogi T, Widiyandari H. Iskandar F electrochemical properties of TiO x/ rGO composite as an electrode for supercapacitors. RSC Advances. 2019;**48**:27896-27903. DOI: 10.1039/ c9ra04346b

[34] Gitashree D, Purna KB, Najrul H, Priyakshree B, Bhagyasmeeta S, Pinaki S, et al. Magnetic nanoparticles towards efficient adsorption of gram positive and gram negative bacteria: An investigation of adsorption parameters and interaction mechanism. Colloids and Surfaces A: Physicochemical and Engineering Aspects. 2017:161-170. DOI: 10.1016/j. colsurfa.2016.12.003

[35] Haopeng C, Cheng B, Xiao H, Wei Q. Synthesis of rGO/p-$Fe_3O_4$@PANI three-phase nanomaterials and electromagnetic wave absorption properties. Materials Research Express. 2019;**6**:125621-125623. DOI: 10.1088/2053-1591/ab5727

[36] Nasiri J, Motamedi E, Naghavi MR. Comparative study of adsorptive role of carbonaceous materials in removal of UV active impurities of paclitaxel extracts. Journal of Pharmaceutical Analysis. 2015;**6**:396-399. DOI: 10.1016/j.jpha. 2015.04.004

[37] Jafari A, Farjami SS, Salouti M, Boustani K. Effect of annealing temperature on magnetic phase transition in $Fe_3O_4$ nanoparticles. Journal of Magnetism and Magnetic Materials. 2015;**379**:305-312. DOI: 10.1016/j. jmmm.2014.12.050

[38] Valenzuela R, Alvarez G, Montiel H, Gutiérrez MP, Mata-Zamora ME,

Barrón F, et al. Characterization of magneticmaterials by low-field microwave absorption techniques. Journal of Magnetism and Magnetic Materials. 2008;**320**:1961-1965. DOI: 10.1016/j.jmmm.2008.02.008

[39] Hafiz SM, Ritikos R, Whitcher TJ, Razib NM, Bien DCS, Chanlek N, et al. A practical carbon dioxide gas sensor using room-temperature hydrogen plasma reduced graphene oxide. Sensors and Actuators B: Chemical. 2014;**193**:692-700. ScholarBank@NUS Repository. DOI: 10.1016/j.snb.2013.12.017

[40] Oliveira AEF, Braga GB, Tarley CRT, et al. Thermally reduced graphene oxide: Synthesis, studies and characterization. Journal of Materials Science. 2018;**53**:12005-12015. DOI: 10.1007/s10853-018-2473-3

[41] Moon I, Lee J, Ruoff R, Lee H. Reduced graphene oxide by chemical graphitization. Nature Communications. 2010;**1**:73-78. DOI: 10.1038/ncomm s1067

[42] Subash P, Shova N, Dipak KG, Ram JY, Amar PY. Synthesis and application of $Fe_3O_4$ nanoparticles in electrochemical pseudocapacitor. Emergent Materials. 2022;**6**:1899-1906. DOI: 10.1007/s42247-022-00417-4

[43] Boucher C, Rubel O, Zhitomirsky I. Supercapacitor performance of magnetite nanoparticles enhanced by a catecholate dispersant: Experiment and theory. Molecules. 2023;**28**(4):1562. DOI: 10.3390/molecules28041562

Chapter 5

# Polyaniline and Its Nanocomposites: From Synthesis to Functional Applications

*Alvena Shahid and Zohra Nazir Kayani*


## Abstract

This chapter discusses the many synthesis methods used to create polyaniline (PANI), one of the most adaptable conducting polymers. It covers traditional methods like chemical oxidative polymerization, aniline dimerization, and electrochemical polymerization, in addition to more complex ones like intercalation and template-assisted polymerization. A discussion of the factors influencing conductivity, molecular weight, and polymerization kinetics is part of the chapter's focus. Particular emphasis is placed on cost-effectiveness and repeatability concerns when scaling up synthesis methods for industrial purposes. The chapter will also look at current advancements in green synthesis, which aim to minimize the material's adverse environmental effects while enhancing its properties for a variety of applications, such as sensors, electronics, and energy storage devices.


**Keywords:** conductive polymers, polyaniline, electrochemical properties, energy storage, supercapacitors

## 1. Introduction

Polyaniline (PANI) has attracted significant attention in recent decades due to its unique combination of environmental stability, electrical conductivity, and ease of synthesis. As one of the most researched conducting polymers, PANI has a diverse range of applications, from electronics to energy storage devices. The ability to modulate its conductivity by changing its oxidation state or through doping makes PANI an excellent candidate for use in flexible electronics, sensors, batteries, and supercapacitors (**Figure 1**) [1].

PANI's remarkable electrical conductivity, robust electronic characteristics, and outstanding processing and modification capabilities have made it the main polymer utilized in recent studies as a novel polymeric material [2]. Flexible electrodes [3], corrosion-resistant coatings [4], lithium-ion batteries [5], supercapacitors [6], fuel cells [7], screen printing [8], sensors [9], and water pollution removal [10] are only a few of the uses for this material. In 1997, MacDiarmid proposed several types of PANI. Fully reduced leucoemeraldine base (LEB), fully oxidized pernigraniline base

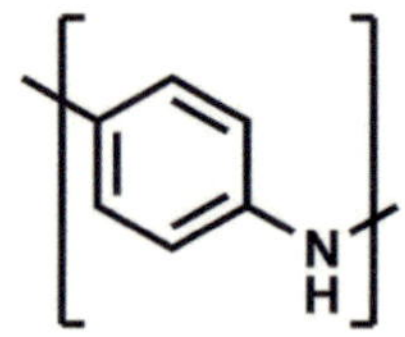

**Figure 1.**
*The monomer of aniline.*

(PAB; **Figure 2**), and partially oxidized emeraldine base (EB) are the three basic, theoretical structures for oxidation states that he postulated.

The type of PANI that has been investigated the most is the EB type. PANI is available in a range of colors. Leucoemeraldine is white or colorless, emeraldine is blue structurally and green as a salt, and pernigraniline varies from blue to violet.

EB can be made conductive by using protonic acids and oxidative doping methods (**Figure 3**); however, this method may not work for LEB and PAB.

**Figure 3**. Basic structure of PANI and different redox forms of PANI with its doped states.

Electrochromic devices and sensors can make use of the color variations associated with PANI in various oxidation states. The method for developing a PANI sensor takes advantage of the significant differences in conductivity between different doping concentrations or oxidation states (**Figure 4**).

### 1.1 Discovery and early development of polyaniline (PANI)

PANI, one of the oldest known conducting polymers, has its origins in the early nineteenth century. The first documented synthesis of PANI dates back to 1834, when the French chemist *Henri R. B. Reichenbach* reported the polymerization of aniline, producing a green-colored material. However, it was not until the 1970s that the true potential of PANI as a conductive material began to be realized [11].

In 1977, the discovery of conducting polymers, mainly *polyacetylene* (another conducting polymer), by *Hideki Shirakawa*, *Alan J. Heeger*, and *Alan G. MacDiarmid* marked a breakthrough in the field of organic electronics. This discovery led to a surge

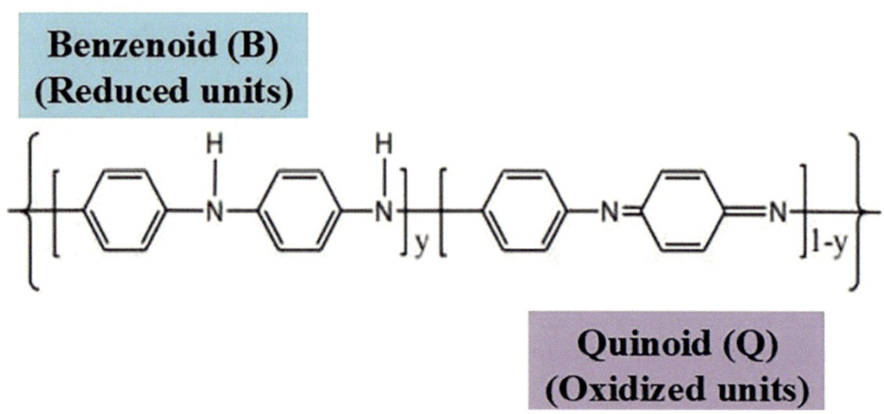


**Figure 2.**
*Molecular structure of polyaniline (PANI).*

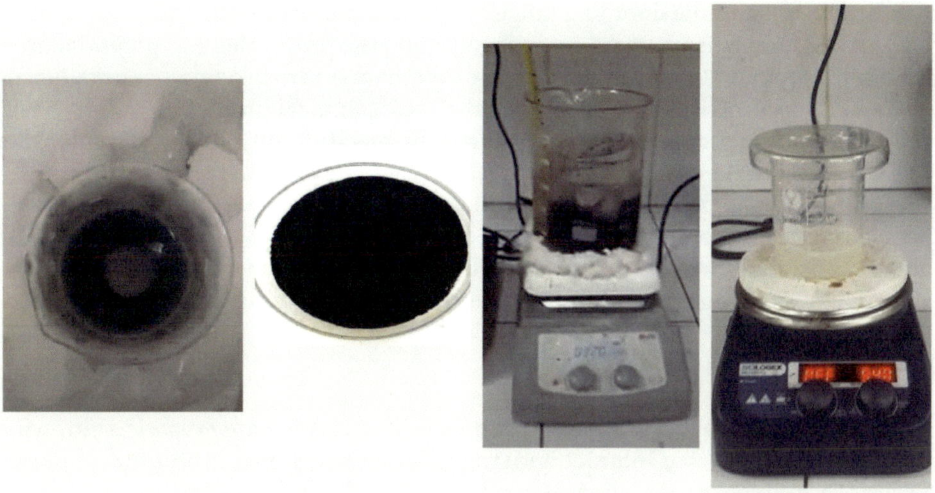

| Emeraldine PANI | powder of emeraldine PANI | pernigraniline PANI | Leucoemeraldine PANI |
|---|---|---|---|

**Figure 3.**
*Different types of polyaniline.*

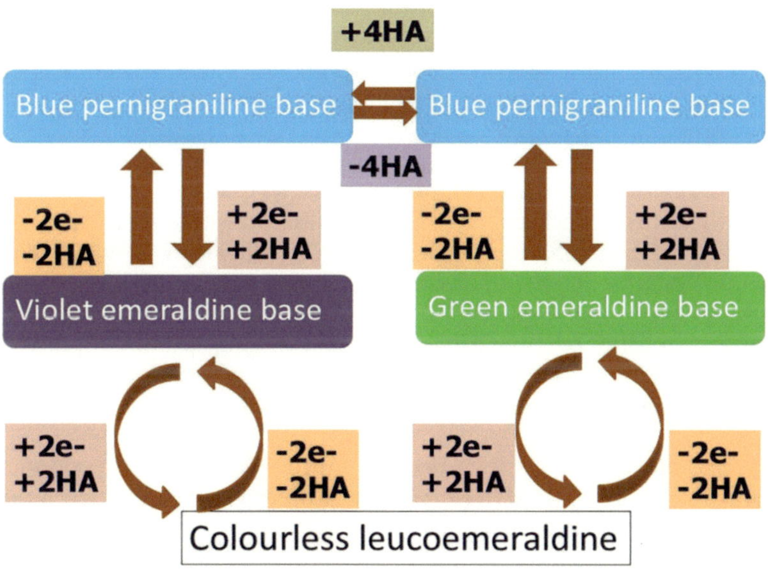


**Figure 4.**
*Salt and base forms of pernigraniline and emeraldine.*

in research on other conducting polymers, including polyaniline, which had previously been overlooked due to its observed insulating properties in its undoped state.

In the 1980s, researchers began to know the conductive behavior of PANI when doped with suitable oxidizing agents. The breakthrough was made by *MacDiarmid and his team*, who showed that PANI when doped with acids such as camphorsulfonic acid (*CSA*) or hydrochloric acid (*HCl*), could show noteworthy electrical conductivity.

This research opened the door to a range of applications for PANI, and since then, broad studies have been devoted to understanding the properties and potential uses of this unique polymer [12]. The main goal of this chapter is to provide a comprehensive overview of the electrical conductivity and electrochemical properties of PANI, with a focus on how these properties can be tailored for specific applications. We will examine the fundamental principles behind PANI's conductivity, the factors that influence its electrochemical behavior, and its practical applications in energy storage and other electronic devices. Through this, we aim to present both theoretical insights and practical considerations for future research and development in this exciting field.

Types of doping agents and their effects on the conductivity of PANI.

### *1.1.1 Protonic acid doping*

The most common doping agents for polyaniline (PANI) are *protonic acids*, which provide protons ($H^+$) that interact with the polymer backbone. Protonic acid doping significantly enhances the electrical conductivity of PANI by converting it from its insulating form (emeraldine base) into the conductive emeraldine salt form. The protonation of the polymer backbone leads to the formation of positively charged sites, allowing for charge delocalization and facilitating electron transport.

- *Common protonic acids used as dopants*:
    - *Hydrochloric acid (HCl)*: One of the most widely used protonic acids for doping PANI. It effectively does the polymer to produce the emeraldine salt form, which is highly conductive. HCl doping can lead to the formation of PANI with good conductivity, but excessive doping may cause stability issues in certain environments.
    - *Camphorsulfonic acid (CSA)*: This dopant is particularly popular for applications where high conductivity and environmental stability are required. CSA-doped PANI exhibits excellent solubility in various solvents, enhancing processability for applications such as thin films and coatings.
    - *Sulfuric acid ($H_2SO_4$)*: Sulfuric acid is often used for producing PANI in its conductive state. It tends to form highly conductive emeraldine salt with improved stability under different environmental conditions.
    - *P-toluenesulfonic acid (pTSA)*: Used to improve solubility and conductivity, this dopant is frequently used in organic electronics and devices that require stable conductivity over time.
- *Effect on conductivity*: The doping of PANI with protonic acids typically increases its electrical conductivity by several orders of magnitude, as the doping process introduces mobile charge carriers (positive polarons or bipolaronic states) in the polymer matrix. The extent of conductivity depends on the concentration of the protonic acid and the extent to which the polymer is doped.

### *1.1.2 Non-protonic (ionic) doping*

In addition to protonic acids, *non-protonic doping agents* (often called ionic dopants) are used to enhance the electrical properties of PANI, typically through the introduction of counter-ions into the polymer. These dopants provide charge balance to the polymer structure and can significantly alter its conductivity, solubility, and mechanical properties.

- *Common non-protonic dopants*:
    - *Tosylate (TOS)*: Tosylate ions are commonly used to dope PANI. They are large, relatively bulky ions that provide a countercharge to the polymer's positive charge in the emeraldine salt form. The use of tosylates can lead to improved conductivity and stability.
    - *Perchlorate ($ClO_4^-$)*: Perchlorate ions are strong oxidizing agents that can provide a high level of doping to PANI, enhancing its conductivity. However, the stability of PANI doped with perchlorate can be sensitive to environmental factors like moisture and temperature.
    - *Polystyrene sulfonate (PSS)*: A well-known polymeric ionic dopant for conductive composites, PSS has multiple advantages, as it contributes to conductivity while also improving the solubility of PANI in other solvents, which is beneficial for developing sensors and electrochromic devices.
- *Effect on conductivity*: Non-amino protonic doping agents induce larger ions, which facilitate PANI charge transport properties without dismantling the architecture of the polymer. These dopants are capable of modifying the geometry and the mechanical properties of PANI, which makes possible the creation of composites with outstanding flexibility and processability.

### *1.1.3 Organic dopants*

Another type of doping agent is the organic dopants that enhance the conductivity of PANI. Engineering-grade dopplers are tailored for specific uses in organic electronics, as for instance, in the fabrication of conducting organic polymer composite materials.

- *Common organic dopants*:
    - *Camphor-10-sulfonic acid (CSA)*: Typically utilized with the expectation of organic and flexible electronics, on account of its capability to both dope and stabilize PANI in different ways. CSA aids in the dissolution of PANI, making it more convenient to process into films or coatings.
    - *Dodecylbenzenesulfonic acid (DBSA)*: An organic compound that enhances the solubility of PANI in organic solvents. It is especially important in the preparation of PANI for use in devices, for example, thin film transistors or organic solar cells.
- *Effect on conductivity*: Because increased charge delocalization takes place with organic dopants, the conductivity of PANI is high. The large organic molecules also enhance the PANI solubility in organic solvents enabling a good dispersion which is a requirement for uniform film or coating formation in flexible electronics and sensors.

### *1.1.4 Metal ion doping*

Doping of metal ions is a different approach to modifying PANI for a particular use, one which seeks to improve its electrochemical activity and charge storage for use in energy devices.

- *Common metal ion dopants*:
  - *Iron ($Fe^{3+}$)*: $Fe^{3+}$ ions are used very often in the doping of PANI for use in supercapacitors and batteries. Iron doping also aids in enhancing the electrochemical outlook of PANI, which, in turn, can improve its realization in energy-storing devices due to the increased activity in redox reactions.
  - *Copper ($Cu^{2+}$) and manganese ($Mn^{2+}$)*: These metal ions can also increase the electrochemical performance of PANI, which makes it applicable in sensors, catalysis, and energy storage devices. Doping with metal ions usually enhances the charge transfer properties and stability of the sensor.
- *Effect on conductivity*: Metal ion doping changes the electrochemical activities of PANI, and in the same way, alters its conductivities. Metal ions are able to do so because they can donate some extra nonbonding electrons to create other redox centers, enhancing the charge transfer and energy-storing ability of PANI devices. This property of metal-doped PANI makes it very applicable in the batteries and supercapacitors.

### *1.1.5 Acid-base doping*

The electrical characteristics of PANI can be modified by using Lewis acids or bases, thus altering it through acid-base doping. This process increases these polymer's stability while modifying their electronic structure.

- *Common acid-base dopants*:
  - *Boron trifluoride ($BF_3$)*: One of the Lewis acid type dopants, boron trifluoride interacts with the nitrogen bases of PANI and modifies its electronic structure so that the conductivity increases.
  - *Ammonium salts (e.g., Tetra-n-butylammonium chloride)*: These ammonium salts can be a base dopant for PANI that limits its charge balance and increases the conductivity.
- *Effect on conductivity*: Stabilization of charge carrier distribution within the polymer enhances its conductivity. Acid-base doping enhances the charge carrier mobility too. This method of doping also has an impact on the solubility and processability of PANI, allowing it to be molded into the necessary shapes for many uses.

## 1.2 Doping effects on conductivity

The charge carriers (polarons, bipolarons, or counter-ions) that are incorporated into PANI when it is doped enable charge transport and greatly enhance its electrical conductivity. The effects of different dopants on the material differ greatly:

- The most widely adopted are protonic acids *HCl and CSA*, which furnish protons required to transform PANI into the conductive salt form known as emeraldine.

- *Non-protonic dopants* (e.g., tosylates, perchlorates) often enhance conductivity by introducing counter-ions.

- *Organic dopants* (e.g., CSA, DBSA) improve solubility and conductivity, making PANI suitable for flexible and organic electronics.

- *Metal ion dopants* (e.g., $Fe^{3+}$, $Cu^{2+}$) enhance the electrochemical properties of PANI for energy storage and sensor applications.

- *Acid-base dopants* (e.g., $BF_3$, ammonium salts) offer another route for improving conductivity and stability.

Each doping strategy provides distinct advantages for specific applications, and ongoing research continues to explore new dopants to optimize PANI's performance in a wide range of devices.

*Electrical conductivity doping effects*: Electrical conductivity will greatly vary depending on the type and concentration of dopants used in PANI. For example, the conductivity was found to increase from \1.98 to 10.2 S/cm when the concentration of the doping HCl was increased from 0 to 1.0 M which caused the formation of polarons in the polymer chain. Also, the use of additives Carbon Nanofibers (CNF) increases conductivity further, to values of 3.73131 S/cm (**Figure 5**) [13].

### 1.3 Electrochemical properties

*Energy storage and conversion*: The redox properties of PANI allow it to be used in applications like rechargeable batteries, supercapacitors, or even fuel cells. Unfortunately, pure PANI lacks stability most of the time, thus voiding their potential use as composites. The enhanced performance of PANI-based composites, particularly those incorporating carbon materials or metal compounds, can be attributed to their superior electrochemical performance due to synergistic effects.

*Industrial applications*: PANI is suitable for the manufacture of passive electronic elements such as resistors and capacitors due to its peculiar phenomena of ionic and electronic conduction. Relying on its promising electrical and electrochemical properties, the challenges of reproducibility and long-term stability put the technology's adoption in glare. Consistency of properties at variations of the synthesis methods and environmental conditions underline the importance of standardized protocols and thorough stability testing. This will hinder broader applications of PANI.

1. *Enhanced conductivity*: The performance of PANI in electrochemical applications has improved due to researchers developing means to enhance the conductivity of PANI using various acids and other dopants.

2. *Supercapacitor applications*: Due to the high theoretical pseudocapacitance of PANI, it has been applied into supercapacitors for a long time. More recent works focused on PANI-based composite electrodes that incorporate PANI with carbon materials or metal oxides to increase the stability and energy density for supercapacitor applications.

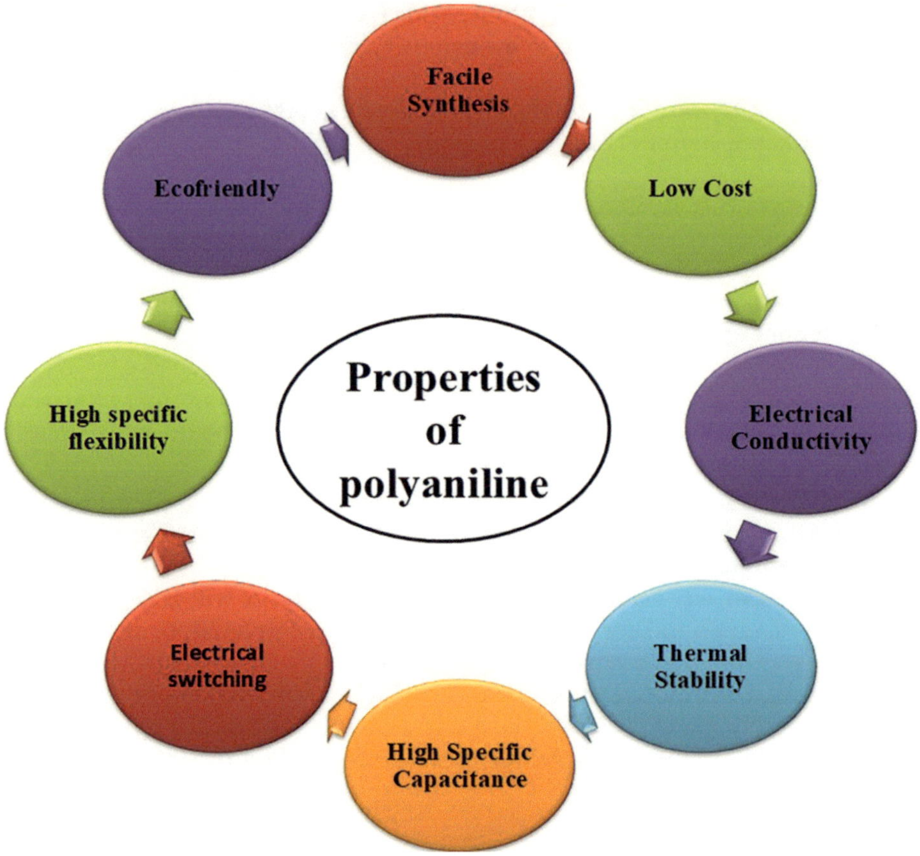


**Figure 5.**
*Properties of polyaniline.*

3. *Composite materials*: The PANI-based nanocomposites exhibit excellent electrochemical performance. PANI-based composites can be grouped into two major categories: binary and ternary.

*PANI-based binary composites*: PANI is known to be combinable with a variety of different types of materials, including metal compounds, carbon materials, and carbon/PANI. In the proceeding part here, we will present the research development on different PANI-based binary composites.

## 2. Synthesis of PANI

PANI is the most appealing electrically conductive polymer due to its easy transition between salt and base forms by the addition of base ($OH^-$) and acid ($H^+$). PANI is one of the top conductive polymer materials because of its special qualities, which include oxidation and reduction, easy fabrication methods, remarkable electrical conductivity, and ecologically friendly stability. PANI has historically been produced by the oxidative polymerization of aniline monomers in an acidic media; however, as research advanced, a number of methods and strategies for its synthesis emerged, including Refs. [14, 15].

- Chemical polymerization.
- Electrochemical polymerization.
- Vapor-phase polymerization (VPP).
- Polymerization employing electron acceptors.
- Photochemically initiated polymerization.
- Enzyme-catalyzed polymerization.

Detailed Research on PANI since its discovery as a conducting polymer has focused on improving its synthesis and properties to satisfy the demand of diverse applications. The synthesis of PANI was initially difficult because of its poor solubility in common solvents and propensity to deteriorate in specific situations. Initially, the synthesis of PANI was somewhat challenging due to its poor solubility in common solvents and its tendency to degrade under certain conditions.

However, in recent years, advances in polymerization techniques have improved the efficiency and versatility of PANI synthesis. Several methods, such as *chemical oxidation polymerization* and *electrochemical polymerization*, have been developed to produce PANI with controlled molecular weights and structures. These advancements have also made it possible to enhance the material's solubility, stability, and conductivity. The key to PANI's unique properties lies in its ability to exist in different oxidation states, from insulating emeraldine base to highly conductive emeraldine salt, depending on the type and extent of doping. This tunability makes PANI especially valuable in applications where changes in electrical conductivity are required [16].

### 2.1 Electrochemical polymerization

The electrochemical polymerization technique is essential for the production of conductive polymers since they are commonly required as large surface area thin films for a range of applications. When conductive polymers are produced using electrochemical polymerization, it becomes challenging to follow the electrochemical production process because research on optical and electrical properties mostly relies on these polymer layers applied to the electrode.

The electro deposition approach for making salt is comparable to this process for making conductive polymers. PANI was prepared using electrochemical process using the galvanic and dynamic approaches.

Furthermore, it is simple to use and make a uniform thin film that is deposited on the glass substrate, the electrochemical method has several advantages over the chemical method [17]. The technique takes place in a basic tube-shaped cell with an electrode, electrolyte solution, and a power supply. The following procedures make up the electro-polymerization technique used to create PANI:

1. Oxidation at the anode that produces a beneficial cationic free radical from aniline monomers

2. By removing protons and rearranging electrons in the aromatic rings, the first stage's structural arrangement produces dimers.
3. The production of new, bigger forms and the appearance of these designed dimers
4. During the last stage, the polymer chain is naturally triggered by the acid in the solution, producing the denatured PANI. This is thought to be the process for PANI's electro-polymerization.

### 2.2 Chemical polymerization

The chemical polymerization method known as oxidative polymerization is used to make polymers. It is believed to be a straightforward, cost-effective approach that may swiftly yield a large range of polymers. For the manufacturing of PANI, this is currently the most widely used and favored approach by many businesses [18]. By introducing a chemical oxidizing agent into the solution, this method initiates the polymerization process and generates an oxidizing force. Oxidizing chemicals employed during the polymerization process, including potassium dichromate ($K_2Cr_2O_7$), sodium vanadate ($NaVO_3$), ammonium persulfate ($(NH_4)_2S_2O_8$), hydrogen peroxide ($H_2O_2$), cerium sulfate ($Ce(SO_4)_2$), and potassium iodate ($KIO_3$). Ammonium persulfate ($(NH_4)_2S_2O_8$) is also one of the most often used oxidizing agents in this process. Ammonium persulfate is commonly used to dissolve aniline, initiate the polymerization process, and stop the formation of unwanted byproducts when aniline is polymerized in an acidic environment ($pH < 3$). The molar ratio of aniline (APS/Aniline) to persulfate is frequently less than 1.2. As was already

|  |  |
| --- | --- |
| Friedlieb Ferdinand Runge (8 February 1794 – 25 March 1867) | Henry Letheby (1816–1876) |

**Figure 6.**
*Early Discoverer of Polyaniline.*

**Figure 7.**
*Discoverer of Polyaniline.*

indicated, the electrochemical polymerization of aniline is similar to the early stages of aniline polymerization with ammonium persulfate. In the first phase, nitrogen electrons are captured by ammonium persulfate to form the radical cation of aniline. The next step involves an interaction between the parasite of another aniline molecule and the radical cations of nitrogen (**Figures 6** and 7). In some cases, the reaction with the radical cation occurs at the para position, which distorts the generated PANI chains. Normally, the radical cation of the resulting aniline molecule and the radical cation of nitrogen in the original aniline compound come into contact at the ortho position.

### 2.3 Vapor-phase polymerization (VPP)

VPP is necessary for PANI their production. This method was initially used in 1998 to coat cotton threads with conductive PANI [19, 20]. The conductive PANI coating was applied to the threads by exposing them to aniline vapor after they had been soaked in ammonium persulfate ($(NH_4)_2S_2O_8$). Polypyrrole, PANI, and polythiophene are among the compounds that can be turned into ultra-thin polymer films using VPP technology. This method, a vapor-phase self-assembly polymerization approach, can be used to produce ultra-pure thin polymer films that are suitable for nanoscale construction, have remarkable superconductivity, and are devoid of impurities.

### 2.4 Photochemically initiated polymerization

An attractive technique for producing PANI with exceptional properties and compositions is photocatalytic polymerization. It also serves as a crucial process for creating hybrid materials. Kobayashi and associates suggested in 1998 and 2001 that PANI could be created by photopolymerization with single or bilayer films, where methyl viologen ($MV^{2+}$) would act as an oxidizing agent and $[Ru(bipy)_3]^{2+}$ would act as a catalyst. PANI would then be produced by exposure to visible light. The triply excited state $^*[Ru(bipy)_3]^{2+}$ is formed when the complex $[Ru(bipy)_3]^{2+}$ is subjected to light with a wavelength of 452 nm. When electrons are moved from $^*[Ru(bipy)_3]^{2+}$ to $MV^{2+}$, the powerful oxidant $[Ru(bipy)_3]^{3+}$is produced. Aniline must undergo this reaction in order to be oxidized and then polymerized into PANI. For potential uses in molecular electronics, the generated PANI was employed in these investigations as monolayer or bilayer films. The irradiation process was then conducted utilizing a variety of methods, including microwaves, gamma rays, X-rays, and ultraviolet light, in order to produce PANI using an external source [21–23].

### 2.5 Enzyme catalyzed polymerization

In previous years, researchers have used oxidizing agents like peroxide to synthesize specific key polymers, such as polypyrrole and PANI, employing enzymes like horseradish peroxidase (HRP) as catalysts [24]. It is believed that this process is environmentally benign because the peroxide is transformed into water; nonetheless, only a few studies have generated branching polymers having tiny molecules.

In order to overcome these difficulties, Samuelson et al. used various electrolytes and templates made of poly (styrene sulfonate) (PSS) in the HRP-catalyzed PANI synthesis solution to produce uniform polymeric chains. Three purposes are fulfilled by PSS in this approach:

1. PSS serves as a structure that aligns the aniline monomers in the correct head-to-tail coupling configuration prior to the polymerization process beginning.

2. It transforms PANI into an electrically conducting emeraldine salt state while providing substantial doping.

3. It gives the (PANI/PSS) molecule water solubility. Water-soluble PANI is produced by this process, which entails polymerization occurring in an acidic environment at pH = 4. The electrolytes that work well are PSS, poly (vinylphosphonic acid), and DNA.

### 2.6 Electrochemical polymerization

According to some, the electrochemical synthesis of PANI is the simplest method because it does not require purification processes that involve separating monomer/initiator molecules and removing solvents. Aniline is polymerized using inert conductive electrodes. Acid and electrolytes are commonly present in water-based solutions for the electrochemical polymerization of aniline [21]. When galvanostatic conditions are present, the maximum current density for aniline polymerization is 10 $mA/cm^2$ [22]. PANI in a range of forms and characteristics can be produced by electrochemical polymerization aniline. A PANI-containing composite is made for application in organic field-effect transistors [23] and rechargeable batteries [10, 11].

### 2.7 Aniline polymerization molds

PANI can be produced in the required configuration through template-directed polymerization by using macromolecular templates such as polyanions, which interact electrostatically with PANI and water-soluble polymers to encourage hydrogen-bonding [25].

This approach simply enhances thermal conductivity by eliminating the high-molecular-mass dopant during its breakdown at high temperatures. PANI can be used to create stable conductive dispersions from low molecular weight acids and water-soluble polymers. During the template polymerization process, aniline is oxidized by ammonium persulfate.

Peroxidases or hemoglobin can also cause aniline polymerization by using hydrogen peroxide and oxygen, whilst laccases play a role in substrate oxidation. Strong polyacids such as poly(2-acrylamido-2-methyl-1-propanesulfonic acid) and poly(4-styrenesulfonic acid) are used in polymerization. Low-molecular-mass acids (templates) and hydrogen-bonding polymers (dopants) are used to produce these forms. Polyvinylpyridine, PANI, polyvinylpyrrolidone, hydroxypropyl cellulose, and polyacrylamide are examples of polymers. Both homogeneous and interfacial environments can benefit from template polymerization. Aniline is dissolved in both the organic (polysalt) and aqueous (anionic surfactant) phases of a solution during interfacial polymerization. PSS-PANI (poly (sodium 4-styrene sulfonate)) and PANI at the interface work together to gradually seep into the liquid. Elongated structures of clustered particles or spherical particles are the results of polymerization of the PANI-polyanion complex [26]. The diameters of the PANI-PSS complex particles depend on the molecular mass of the polyanion and the concentration of the homogeneous template polymerization. When the molecular weight is greater than 6800 PSS-Na, the interpolymer complex has a particle diameter of 2–3 nm at [aniline] = [PSS-Na] $4.5 \times 10^{-2}$ mol/L. Aggregates of fibrillar material between 70 and 90 nm and 200 and 1 nm in size will occur if the component concentration rises above this limit. Polymerized PSS-Na particles range in size from 100 to 130 nm and can aggregate into smaller particles (**Figure 8**).

The size and shape of PANIs formed in hydrogen-bonding polymers at low aniline concentrations ($4.5 \times 10^{-2}$ mol/L) differ based on the particular polymer and dopant employed. Particles produced by polyvinylpyrrolidone-hydrochloric acid aniline template polymerization range in size from 50 to 150 nm. Aniline reacts with hydrochloric acid and poly(vinyl alcohol) to form PANI structures that are 200–300 nm

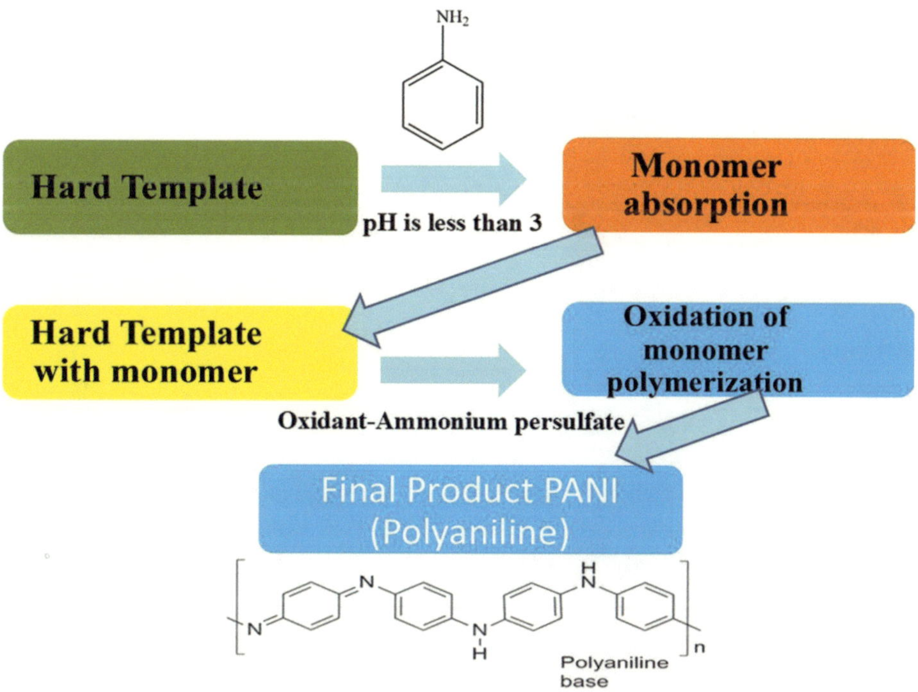


**Figure 8.**
*Polymerization of aniline.*

wide. Particles measuring 300 nm are formed by interpolymer complexes when hydrochloric acid is replaced with p-toluenesulfonic acid [27].

The ratio of aniline to polyelectrolyte templates in the process environment influences phase-separable PANI-polyelectrolyte complexes or soluble in water nonstoichiometric complexes. The ratio of aniline to PSS-Na in the flakes produced by the facility is 4:1. The protonated nitrogen atoms in PANI's Quinonediimine balance the template charges due to its nitrogen-to-sulfur ratio of 2.3. The type or concentration of polyacid has no bearing on the complicated conductivity of template synthesis. Composite materials made using real polymer soft templates can be altered. PANI-PSS fibers have the ability to promote the deposition of ruthenium and platinum for the oxidation process of methanol.

## 3. Properties

### 3.1 Electrical and doping properties of PANI

PANI is a highly attractive conductive polymer that is frequently used in the electrical industry due to its outstanding electrical conductivity in the emerald salt state. By protonation of the nitrogen in the imine group and converting it to a salt, acid transforms the basic emeraldine form of PANI from a nonconductive to a conductive state. For undoped base emeraldine, PANI's conductivity typically ranges from $\sigma < 10–10$ S/cm, while for doped salt emeraldine, it is $\sigma > 10$ S/cm [27]. The method for rapidly raising the electrical conductivity or altering other PANI properties is connected to the doping of the polymer.

In recent years, many researchers across the world have focused on doping PANI with different compounds for use in a range of applications. The main goal of this is to modify the electrical conductivity of the newly produced compounds (polymer compounds) for usage in that field. **Table 1** displays the electrical conductivity of the materials produced *via* PANI.

| Solvent/dopant (0.1 M) | Color change time (min) | Weight of produced polymer (g) | Electrical conductivity (S/cm) |
|---|---|---|---|
| $H_2O$/HCl | 4 | 0.20 | 89 |
| Ethanol/HCl | 4 | 0.07 | 22 |
| $CH_3CN$/HCl | 6 | 0.09 | 14 |
| THF/HCl | 5 | 0.15 | 14 |
| DMSO/HCl | 6 | 0.17 | 15 |
| *m*-Cresol/HCl | 4 | 0.20 | 11 |
| Acetone/HCl | 10 | 0.18 | 11 |
| $CH_2Cl_2$/PTSA | 45 | 0.11 | 0.001 |
| Benzene/PTSA | 45 | 0.11 | 0.003 |
| Nitrobenzene/PTSA | 15 | 0.05 | 0.042 |
| Cyclohexane/PTSA | 60 | 0.12 | 0.0004 |

**Table 1.**
*Conductivities of conductive polymers with selected dopants.*

The electrical conductivity of polyaniline prepared using different solvents and $(NH_4)_2S_2O_8$ as an initiator is displayed in **Table 1** for a duration of 45 minutes in an electric field of 10 $kV/cm^2$. The first monomer, aniline, has a weight of 0.1 g.

### 3.2 Semiconducting properties

PANI is one of the most advantageous materials in electronics due to its exceptional electron conductivity. Yakuphanoglu [28] and ÿenkal developed the PANI diode, and the PANI ionic liquid's electrical and thermal conductivity experiments revealed that this polymer functions as a p-type semiconductor. The electrical properties of PANI films that had been infused with boric acid, on the other hand, were investigated. Studies showed that PANI films with boric acid have semi-conductive qualities; their electrical conductivity was $1.02 \times 10^{-4}$ S/cm at ambient temperature. Boric acid-doped PANI films were found to have an optical band gap of 3.71 eV. PANI is widely utilized in organic field-effect transistors and solar batteries because of these semiconductor properties.

### 3.3 Capacitive properties

In recent years, PANI has been widely used in many electrical and electronic applications, particularly in energy storage devices such as supercapacitors. Due to the rapid redox reactions occurring in various regions of the polymer material, PANI has contributed to the improvement of these energy-efficient devices' properties. PANI offers better capacitive responsiveness and greater specific energies than double-layer capacitors. In addition, PANI has a considerable electrical conductivity and a greater effect than inorganic batteries. However, throughout the charge-discharge cycle, pure PANI capacitors undergo contractions. Researchers investigated blending various compounds with PANI to address these problems and enhance the capacitors' properties. These compounds included PANI–carbon nanostructures [25], PANI–graphene [16, 29], PANI–Cerium Oxide [6], PANI–silver–manganese dioxide [30], and PANI/$MoS_2 - MnO_2$ [31], among others. Significant gains in stability, lifetime, and energy density were demonstrated by the PANI-assisted electrode changes.

## 4. Significance of PANI

Polyaniline holds a central place in the family of conducting polymers due to its remarkable properties and versatility. When compared to other conducting polymers, such as polypyrrole or polythiophene, PANI stands out for its ease of synthesis, low cost, and environmental stability. Its conductivity can be easily tuned by varying the doping level or oxidation state, which makes it an ideal candidate for various applications in electronics, energy storage, sensors, and environmental monitoring.

Another advantage of PANI over other conducting polymers is its environmental stability. Many conducting polymers tend to degrade when exposed to light, oxygen, or humidity, but PANI exhibits good stability under a wide range of environmental conditions. This makes it suitable for long-term use in practical devices, such as organic light-emitting diodes (OLEDs), batteries, supercapacitors, and electrochemical sensors.

Moreover, PANI's ability to undergo reversible redox transitions adds a layer of versatility, especially in electrochemical applications. The polymer can switch between insulating and conducting states depending on the charge state, which allows it to function in energy storage devices, such as rechargeable batteries and capacitors.

The field of conducting polymers, and particularly PANI, has witnessed significant progress over the past few decades. Researchers continue to explore new doping strategies, composites with other nanomaterials (such as carbon nanotubes and graphene), and applications in energy harvesting, wearable electronics, and bioelectronics. With the rise of flexible electronics and energy storage systems, PANI's unique combination of conductivity, processability, and environmental stability positions it as a critical material in the development of next-generation technologies.

The electrical conductivity of PANI is one of its most defining characteristics. Unlike traditional metals or semiconductors, conductive polymers like PANI offer advantages such as light weight, flexibility, and low production costs. The conductivity of PANI can be tuned by controlling the degree of doping or by varying the polymer's oxidation state, making it a versatile material for a wide range of electronic applications. In this chapter, we will explore the factors that influence the electrical conductivity of PANI, including the role of doping agents, the impact of environmental factors, and the molecular structure of the polymer itself.

In addition to its electrical conductivity, PANI's electrochemical properties play a crucial role in determining its performance in various devices. The polymer undergoes reversible redox transitions between different oxidation states, which significantly affect its electrical conductivity and charge storage capabilities. Understanding these electrochemical behaviors is essential for optimizing PANI for energy-related applications, such as rechargeable batteries and capacitors. This chapter will also delve into the mechanisms behind the electrochemical processes in PANI, highlighting key electrochemical techniques such as cyclic voltammetry and impedance spectroscopy used to study its behavior.

## 5. Composites

### 5.1 PANI/carbon binary composites

PANI/carbon binary composites have great potential for enhanced performance in supercapacitor electrodes due to their remarkable electrical properties and stability. It has been shown that porous carbon, graphene, carbon nanofibers, and carbon nanotubes all have remarkable qualities in addition to PANI. Materials made of porous carbon have exceptional stability, great porosity, and high specific surface area. Chen et al. produced PANI using the electrochemical polymerization process and combined it with porous carbon electrodes to demonstrate an initial specific capacitance of 180 F g – 1, which is almost twice as high as the bare-carbon capacitor (BC). Zhang et al. synthesized highly porous activated carbon composites based on polyaniline/cellulose (PANI/C-ACs) using a straightforward and affordable method. These composites significantly outperformed pure PANI in terms of specific capacitance (765 F g – 1 at 1 A/g) and cycling stability (91% capacitance retention after 5000 cycles). Ordered mesoporous carbon (OMC) is a type of porous carbon that has a higher specific surface area (1000–2000 $m^2/g$) than ordinary porous carbon. The exceptional faradaic capacitance of PANI suggests that PANI/OMC composites have a lot of promise for improving supercapacitor performance. The advantages of PANI/OMC-based composites for supercapacitor electrodes have been the focus of several important studies. These composites are effectively produced by chemical polymerization or *in situ* polymerization. Another type of porous carbon material is carbon sphere-type materials. Because of their huge specific surface area and pore topologies, which improve their capacity to move electrons from the electrolyte to the electrode surface in supercapacitors, these materials are very important.

### 5.2 PANI and carbon nanotube (CNT) composite

Multiwall nanotubes (MWNT) and single-walled nanotubes (SWNT) are the two main categories of carbon nanotubes (CNTs) [13]. In addition to having a large surface area, CNTs have remarkable electrical, mechanical, and thermal properties. Over the years, the PANI/CNT composite has been explored as an electrode for supercapacitors to increase electrochemical performance. Solution mixing, electrophoretic, electrochemical, interfacial, and *in situ* chemical polymerization have all been used to produce PANI-CNT. Studies have indicated that the addition of carbon nanotubes (CNT) to the PANI-CNT composite enhanced the composite's electrical, polymerization, shape, and crystallization. The PANI-CNT nanocomposite, which is created by electrochemical polymerization in which PANI is coated onto CNT, has enhanced mechanical stability, electrical conductivity, and electrochemical properties. When the solution mixing method was used to form the composite, CNT clumping was much less frequent. Gupta et al. used electrochemical polymerization to make PANI/SWCNT composite electrodes with varying weight percent compositions. For electrochemical characterization, a three-electrode setup using an electrolyte of sulfuric acid was employed; Pt served as the counter electrode and saturated calomel electrode (SCE) as the reference electrode. Microstructural studies demonstrated that the PANI coating covers the single-walled carbon nanotube (SWCNT), facilitating the rise in the composite's electrical conductivity. According to electrochemical analyzes, 73% PANI/SWCNT achieved the maximum specific capacitance, energy density, and power density, which were 228 Wh kg − 1, 2250 W kg − 1, and 485 F g − 1, respectively. Ninety four percent of the capacity was retained after 1500 cycles. The electrochemical characteristics of PANI nanofibers and PANI/CNT composite, which were made *via* interfacial polymerization and *in situ* chemical polymerization, respectively, were examined by Sivakkumar et al. Using 1.0 M sulfuric acid, the electrochemical behavior was examined in a two-electrode setup. According to the results, the specific capacitance of PANI nanofiber was 554 F g − 1 at 1.0 A g − 1, which is almost 10 times greater than the capacitance attained using a traditional macroscopic PANI electrode. The large specific surface area of PANI nanofibers, which facilitates energy storage through double-layer and Faradaic mechanisms, was associated with the high specific capacitance value attained. After 1000 cycles, however, the PANI nanofiber's capacitance decreased to 57 F g − 1, indicating a capacity retention of roughly 10.3%. The study noted that the material's expansion and contraction caused significant stress on the PANI backbone, which in turn led to a decline in PANI's electrochemical performance. Furthermore, during the electrode's cycles of charging and discharging, the addition of CNT to the polymer improved PANI's mechanical stability and electrical properties. After 1000 cycles, PANI/CNT demonstrated a capacity retention of 63.7% and a specific capacitance of 606 F g − 1 at 1.0 A g− 1. The results showed that CNT fillers are still a good option for improving PANI composites' performance in energy storage applications. Yan and his associates discovered that the mechanical characteristics of PANI/GNS composites were significantly improved by the addition of 1 weight percent CNT. The study found that adding carbon nanotubes (CNT) to the electrode significantly reduced the rate at which the composite material was destroyed during the doping and dedoping procedures. Research indicates that PANI/GNS composites without CNT exhibited a specific capacitance of 1046 F g– 1 and retained a 52% retention capacity, while PANI/GNS/CNT-1 wt% achieved 1035 F g – 1 and a 94% retention capacity after 1000 cycles. Another study by Zhang et al. used mechanical mixing to form a multi-walled carbon nanotube/polyaniline (MWCNT/PANI) nanocomposite, which was then cemented *via* spark sintering. The results demonstrated that a small amount of MWCNT improved the composite's electrical conductivity, reaching a peak conductivity of 159 S m – 1 at 30% MWCNT. Furthermore, adding MWCNT had no effect on the material's thermal conductivity. Using a one-step electrospinning process, Simo two et al. created nanostructured polyaniline-carbon nanotube supercapacitor electrodes. For the investigation, a 93% PANI and 7% polyethylene oxide (PEO) mixture was created. The main goal of the PEO was to encourage suitable chain entanglement in order to produce the electrode using electrospinning. PANI/CNT/PEO in different weight percentages of 93, 7, and 12% were also synthesized and investigated. The electrochemical properties were evaluated using GCD, CV, and electrochemical impedance spectroscopy (EIS). The results showed that PANI-CNT-PEO and PANI/PEO, respectively, produced specific capacitances of 308 and 385 F g – 1. Following 1000 cycles, the capacity retention rates for PANI-CNT-PEO and PANI/PEO were 70% and 81.4%, respectively. These composites often serve as networks and conductive layers that boost the overall efficiency of the energy storage parts of supercapacitors. The research recommends that synthesizing PANI/carbon binary composites can enhance electrochemical performance. The morphology and structure of these composites considerably improve the electrochemical behavior of PANI-based electrodes. Moreover, carbon-related materials still struggle with the improvement of specific capacitance. Metal compounds, mainly metal oxides, can help to address this main issue, making the research of PANI/metal compounds suitable for supercapacitors.

## 5.3 PANI-based ternary composites

Supercapacitor electrodes have benefited greatly from PANI-based binary composites; nonetheless, their ideal electrochemical properties, such as conductivity, specific capacitance, and cycling stability, have not yet been attained. Ternary composites are becoming increasingly important for achieving the optimum electrochemical performance. Because of their exceptional performance as active electrode materials, ternary composites of PANI, metal oxides, and carbon are primarily used in supercapacitors. In comparison to the other samples, the CMP3 composite, which contains 40% PANI, showed the lowest band gap energy at 2.19 eV and an extraordinarily high specific capacitance of 2247.962 Fg – 1 [25]. The material demonstrated outstanding electrochemical properties, including notable specific capacitance and outstanding cycling stability. Using 3D polystyrene microsphere-reduced graphene oxide/MnO2/PANI (3DrGN-MnO2-PANI), Fu's group produced a coaxial array composite that demonstrated exceptional cycling stability and improved specific capacitance. The ternary porous hierarchical PANI/RGO/$CeO_2$ hybrid microsphere that Jeyaranjan and his group developed is renowned for its remarkable rate capability, outstanding long-term cycling stability, and superior specific capacitance. The three components' advantageous and cooperative interaction results in enhanced electrochemical properties. PANI/metal oxides/carbon compounds refer to a kind of ternary composites composed of PANI that are utilized in supercapacitor electrodes. Further components, such as Ag/MnO2/PANI nanocomposite films, can enhance the strength of the synergistic effects. These composites enhance electrochemical performance *via* exceptional electronic conductivity, rapid ion transfer, stable vermicular shape, and a clustered vermicular-like structure. Composites of cobalt oxide: The (Co3O4/rGO)/PANI (60%:40%) composite showed remarkable cycling stability with 92% retention after 10,000 cycles, achieving a specific capacitance of 3103.27 F/g at 2 A/g [29]. In order to create hierarchical ZnO@metal-organic framework (MOF)@PANI core-shell nanorod arrays for use as a supercapacitor electrode, Zhu and his colleagues devised a technique that achieved high specific capacitance, exceptional rate performance, and extended cycling stability on carbon cloth. These composites' exceptional hierarchical core-shell structure and the cooperative interaction of the three essential constituents are linked to their enhanced electrochemical properties.

### 5.4 PANI-based quaternary composites

Using a simple hydrothermal and *in-situ* polymerization method, Shahid and their colleagues created $Pr_2CuO_4$/MXene/MWCNT and PANI composites for symmetrical supercapacitor applications. $Pr_2CuO_4$ exhibited a tetragonal single phase in the composites, with the (103) plane showing a preferred orientation peak. The dielectric constant was the highest and the absorption range was the widest for the PC2 composite. The enhanced PC2 electrode achieved a volumetric and gravimetric capacitance of 2611.47 $Fg^{-1}$ and showed outstanding electrochemical performance at a current density of 2 Ag − 1. Environmental sustainability: PANI is prized for its adaptability, user-friendliness, and positive environmental effects. Its redox characteristics and resistance to corrosion make it a valuable material for a variety of applications. Conductive polymers are becoming more and more popular as materials for energy storage because of their remarkable qualities [28, 29]. In their early form, conducting polymers act as insulators, with conductivity rates between $10^{-10}$ and $10^{-5}$ S $cm^{-1}$ [30]. Additionally, the doping method, which employs either positive or negative dopants, can raise their electrical conductivity. This results in materials with exceptional conductivity levels between 1 and $10^4$ S $cm^{-1}$ [31]. Conductive polymers' appropriateness for the creation and invention of energy storage systems is greatly enhanced by modifying the doping and dedoping procedures. Among the various applications of conductive polymers are supercapacitors [13], batteries [30], sensors [30], solar cells [32], actuators, electrochromic devices, fuel cells, sensing devices, photodynamic therapy, and bioimaging. Additionally, conductive polymers offer high specific capacitance due to their potential to absorb charge throughout the redox process and store it along the entire molecular chain. Electrical conductivity, rate capability, and power density are all enhanced when conductive polymers are utilized in place of inorganic metal oxides. Compared to carbon-based supercapacitors, conducting polymer-based energy storage electrodes have a greater energy density [25]. Polyaniline (PANI) is the most frequently utilized conducting polymer in tissue engineering, drug delivery, supercapacitors, fuel cells, neural electrodes, batteries, and tissue sensors. PANI's advantageous electrochemical properties, quick and reversible redox reactions, wide voltage ranges, good conductivity when doped, ease of manufacturing, affordability, and low environmental impact are the main reasons why researchers and companies are interested in using it as an energy storage material [21, 28]. Furthermore, PANI in its initial condition experiences volumetric changes (swelling, cracking, and shrinking) throughout the redox process when used as an electrode material in energy storage devices, which leads to mechanical degradation and capacity reduction [21]. Recent engineering and research efforts have concentrated on incorporating filler materials into the conducting polymer matrix to produce composites in order to overcome these constraints. These conductive polymer composites have better properties than pure PANI because of the combined impact of the basic materials [21]. According to research, PANI's mechanical, electrical, and stability properties are enhanced by the addition of inorganic ceramics or carbon elements. Polyaniline (PANI) is one polymer that is well-known for its remarkable electrical conductivity and electrochemical properties. Although it may vary based on the synthesis process, PANI's electrical conductivity usually falls between 2 and 10 S/cm when employing the pressed-pellet method. When emeraldine is treated with acid, its conductivity may rise by roughly 10 orders of magnitude. Some research claims that PANI's electrochemical qualities were applied to supercapacitors. These included an energy density of 8.88 kW/kg, a power density of 96 W h/kg, a specific capacitance of 503 F/g, and an 85% cycling stability over 10,000 cycles. PANI was used as a gas sensor because of its exceptional electrical conductivity and capacity to generate a sizable surface area. The movement of ionic and electronic charges between electrodes and biological systems can be managed *via* interfacial charge transfer PANI. Three oxidation forms of PANI are leucoemeraldine (white/clear), pernigraniline (blue/violet), and emeraldine (salt-green/base-blue). The most well-known and conductive PANI kind is emeraldine. The instability and poor solubility of the emeraldine base form of PANI at melt-processing temperatures make it difficult to synthesize. Combining PANI with biopolymers can lessen its disadvantages.

PANI nanocomposites: Any novel material created by mixing two or more constituents with radically differing properties is called a composite. The newly created composites exhibit a number of traits or improved properties in comparison to the original materials. In order to reveal innovative, diverse, and highly advanced qualities, the comprehension requires creating a new composite. Notwithstanding PANI's outstanding attributes and alluring characteristics, many applications have limitations. PANI needs to be combined with other materials to create polymeric composites in order to boost its effectiveness in a range of applications. When the included material is at the nanoscale, the final product is referred to as a polymeric nanocomposite. This is because the development of combining materials with polymers produces polymeric composites. It has been observed that PANI composites were created by combining PANI with different materials. Metals, carbon nanostructures, metal oxides, polymers, and various organic compounds are some examples of the most recent additions used to PANI for the creation of PANI composites. **Table 2** displays the variety of PANI nanocomposites mixed with metal nanoparticles. PANI

| Electrode materials | Current density (or) scan rate | Specific capacitance (F/g) | Cyclic stability | Ref. |
|---|---|---|---|---|
| ($CeO_2$/MXene)/PANI (40%:60%) | 2 $Ag^{-1}$ | 2247.962 | 91.7% After 6000 cycles | [25] |
| PANI/MoOx nanowires decorated MXene | 5 mV/s | 450 | 91.5 After 10,000 | [32] |
| ($Pr_2CuO_4$/ MWCNT/ MXene): PANI 50:50 (PC2) | 2 $Ag^{-1}$ | 2611.47 F/g | 93.95% After 10,000 cycles | [6] |

**Table 2.**
*Comparison of the specific capacitance performance of metal oxide@PANI-based supercapacitor electrodes.*

| PANI nanocomposite type | Dopant | Preparation method | References |
|---|---|---|---|
| PANI/SWCNT | SWCNT | Oxidative polymerization method | [33] |
| PANI/MWCNT | MWCNT | Oxidative polymerization method | [34] |
| MWCNTs/PANI/$TiO_2$ | MWCNT/$TiO_2$ | One-pot polymerization process | [35] |
| PANI/GO | GO | *In situ* polymerization technique | [16, 36] |
| PANI/graphene | G | *In situ* polymerization | [37] |
| Polyaniline/graphite oxide | Graphite oxide | Interfacial polymerization | [38] |
| Polyaniline/fullerene $C_{60}$ | $C_{60}$ | Copolymerization of aniline | [39] |
| PANI-RGO-ZnO | RGO-ZnO | *In situ* oxidative polymerization | [40] |

**Table 3.**
*An overview of the production of several polyaniline nanocomposites based on carbon nanocompounds.*

composites have been developed employing metallic nano-oxides due to its significance in a range of applications, such as electrical storage systems, sensors, supercapacitors, water purification, solar cells, and electronic devices.

However, due to the high strength of the resultant nanocomposite materials, PANI was commonly combined with carbon nanocompounds. The resulting PANI nanocomposites have demonstrated remarkable efficacy in their respective roles and have been used in a wide range of applications. Different PANI nanocomposites containing different carbon nanocompounds are shown in **Table 3**. The relevance of PANI and its wide range of applications in both the experimental and engineering domains are further highlighted by the numerous polymeric nanocomposites of PANI that are accessible in a variety of shapes and configurations, utilizing a number of applications and polymerization techniques.

## 6. Applications of PANI

Conductive polymers have shown special qualities since their invention because of their distinct physical, chemical, and electrical configurations, which make them appropriate materials for a range of applications. For instance, PANI is a conductive polymer that has piqued the interest of numerous researchers for a variety of uses due to its exceptional electrical characteristics, affordability, ease of synthesis, distinctive features, and environmental friendliness. The following describes the main applications of PANI, either as primary or additional content.

### 6.1 In supercapacitor

Fossil fuels, natural gas, and gasoline are just a few of the energy sources that have been used extensively due to the global economy. These resources have a damaging effect on the environment even if they provide a lot of energy. Alternative energy sources must thus be used to replace the traditional energy system. Scientists have been working to develop and improve high energy, sustainable, and renewable energy sources such fuel cells, wind, hydropower, and supercapacitors in order to achieve this.

Supercapacitors are widely used in many different applications, such as electrical equipment, electronics, and wearable technology, making them one of the most promising energy solutions with a large market potential. The main way that supercapacitors differ from traditional capacitors is that they can store hundreds of times more energy than dielectric capacitors, in addition to having a higher energy density, a longer lifespan, and faster charging and discharging rates. Supercapacitors can be divided into three main categories:

(i) Double-layer electrostatic capacitors usually use carbon-based electrodes like carbon, graphene, and carbon nanotubes (CNTs); (ii) pseudo-electrochemical capacitors, which mainly use conductive polymers, metal sulfides, and metal oxides as their electrode materials; and (iii) hybrid pseudocapacitors, which combine the features of the first two categories. The materials utilized for the electrodes in supercapacitors have a significant impact on the energy storage device's efficiency and energy conversion mechanism. The electrodes in supercapacitor devices include carbon, conductive polymers, and metals, depending on their specific use.

Supercapacitors made of various carbon materials provide remarkable conductivity and durability despite their high energy density, which is still relatively low. Metals have particularly advantageous electrochemical properties when used in supercapacitors, but their use in this sector is limited by their natural occurrence, extraction challenges, and expensive cost. PANI has been effectively used in supercapacitors due to its remarkable conductivity, specific capacitance, and spectrum of oxidation states. Furthermore, PANI is perfect for significant applications in the manufacturing of supercapacitor electrodes due to its varied electrochemical properties.

The electrochemical efficiency of supercapacitors that incorporate PANI as the principal electrode material depends on the inherent qualities of PANI, such as its chemical and physical characteristics, production process, dopant type, and nanostructure [41]. PANI has been combined with graphene, fullerene, graphene oxide, and carbon nanotubes, among other carbon materials, to generate supercapacitors. Metal nanoparticles have also been employed in conjunction with PANI to achieve the same goal.

### 6.2 In sensors

A sensor is a device that can identify various changes in its environment and ascertain whether other objects are there. Nowadays, inorganic, organic, or semiconducting materials make up common sensors. The need for conductive polymers as a material for sensor development has significantly increased recently. Many morphologies and topologies, such as microstructures, nanowires, nanotubes, nanoparticles, and nanosheets, have made PANI a polymer of great interest for use as sensors. PANI or its nanocomposites have been used to build a variety of sensors, including chemical and biological ones.

Researchers have used a variety of nanostructures, such as nanofibers, nanowires, and nanotubes, to develop PANI. Because of their superior surface area and gas-release capabilities, these nanostructures have since been used in gas sensors. Since the gas molecules do not change PANI's conductivity, unmodified PANI cannot detect $H_2S$ gas on its own. Thus, PANI and gold nanoparticles are combined to generate gold nanocomposites, which allow for effective $H_2S$ gas detection with response rates even at extremely low concentrations (0.1 ppb). PANI nanostructures were employed to detect methanol, carbon oxides, ammonia, hydrogen chloride, hydrazine, and chloroform. This has led to a substantial structural change in the chains of PANI nanostructures [38]. By combining several nanoparticles to form PANI nanocomposites, the sensing capabilities of PANI can be improved. This makes it possible to detect gases or compounds at incredibly low concentrations and reduces the sensing time (**Figure 9**).

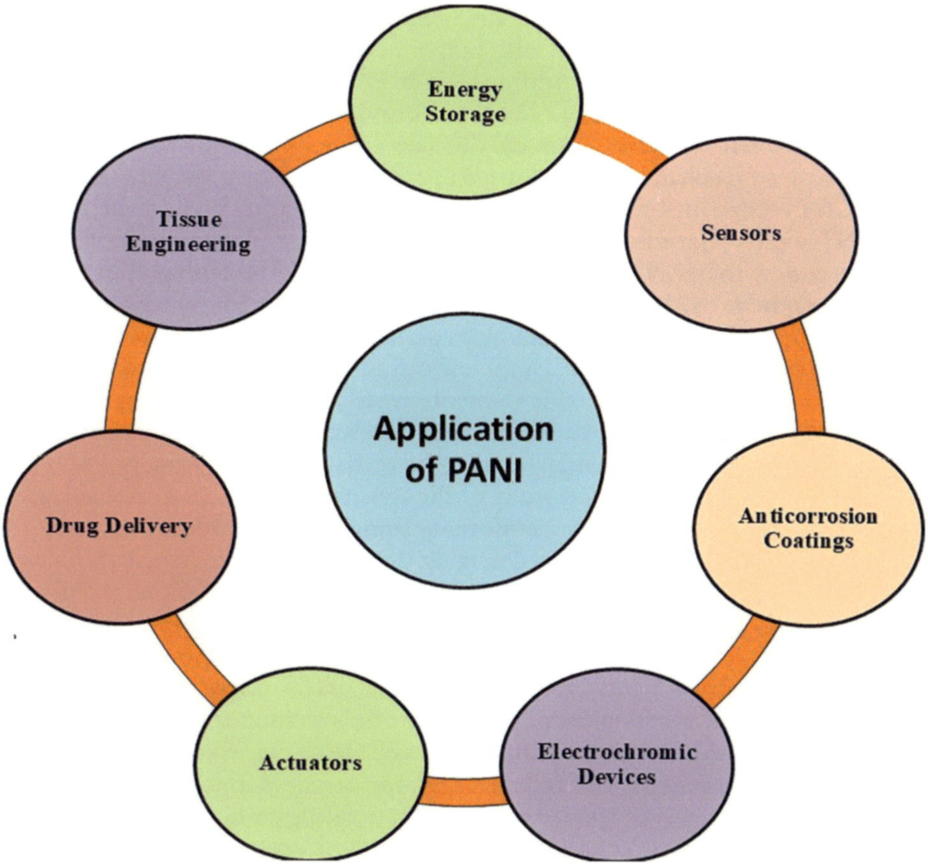


**Figure 9.**
*Applications of polyaniline.*

### 6.3 The challenges facing PANI and its composites as energy storage material and recommendation

PANI is a conductive polymer with electrochemical properties that exhibits exceptional pseudocapacitive energy storage properties, is environmentally benign, and is easily manufactured without sacrificing flexibility. PANI has efficient electron conductivity, which permits a Faradaic reaction to store energy in its material mass. However, PANI's use as an energy storage medium has been limited by its poor mechanical and chemical stability throughout the charge-discharge cycle.

According to studies, PANI produced at the nanoscale enhances ionic and electron mobility in the electrode material, expands the number of redox reaction sites, and lessens the structural degradation of the polymer during charge-discharge cycles. Numerous studies have been carried out with the goal of using various processing methods to create PANI in a range of sizes and forms, including nanofibers, nanorods, nanospheres, and nanowires. However, from the standpoint of nanofillers, the review study focused primarily on the most recent advancements in PANI composites for energy storage applications. It has been shown in the review of different PANI nanocomposites that PANI composites that include electric double-layer carbon-based reinforcing elements, like graphene and carbon nanotubes, are more effective at addressing the issues of reduced cyclability and chemical instability associated with conducting polymers. PANI treated with graphene and/or carbon nanotubes exhibits improved electrical conductivity. Adding CNT to PANI may result in higher porosity, a more uniform distribution of pore sizes, and provide the electrode with the structure required for efficient electrolyte ion diffusion. The PANI/CNT nanocomposite exhibits enhanced mechanical and thermal properties. However, researchers working on PANI/CNT composites continue to confront difficulties due to the small surface area of CNT and its poor dispersion inside the PANI matrix. A composite with a larger surface area than CNT is produced by augmenting PANI with graphene oxide or reduced graphene oxide, which is a desirable characteristic for materials utilized in energy storage. However, graphene's aggregation, stacking, and restacking have shown to be serious disadvantages. Graphene can be used as a spacer and is less likely to cluster and layer when metal oxide elements are incorporated utilizing the proper processing techniques. This improves the composite's overall performance as an electrode for a supercapacitor. Furthermore, the actual manufacturing of PANI-based supercapacitors requires a detailed study of the structural, interfacial, and surface interactions between PANI and the reinforcing components in order to achieve maximum performance. By carefully changing the ratios of composite materials, the supercapacitor electrode may achieve the necessary performance, which is defined by enhanced energy, power density, and rate capability. Further study into the development of a solid-state supercapacitor system using a PANI-based nanocomposite with a broad voltage spectrum is crucial to improving the performance and usability of supercapacitors. It is important to consider a synthesis technique that ensures low internal resistance, efficient movement of ions and electrons, a large contact surface area, sufficient homogeneity, and reduced agglomeration. Methods like solution mixing and heat processing seem to work well for creating conducting polymer-based nanocomposites. On the other hand, mixing solutions is very cost-effective and gives you total control over the composite's constituent parts, ensuring that the reinforcing phases are distributed evenly and effectively throughout the polymer matrix. Crosslinking and notable improvements in the electrode material's mechanical, chemical, and thermal stability will result from the polymer composite's annealing procedure. Using computer tools also necessitates careful study of the properties and functions of supercapacitor electrodes during the charging, storage, and discharging phases. To increase the efficiency of energy storage, these computational methods should focus on the molecular dynamics and interactions related to the electrolyte and electrode material.

## 7. Conclusions and outlook

Ongoing research aims to further improve the electrical and electrochemical properties of PANI composites. This includes exploring new dopants, reinforcing materials, and advanced processing techniques.

These breakthroughs highlight the potential of PANI in advancing sustainable energy storage solutions and improving the performance of electrochemical devices. Incorporating these key points can underscore the significance of the chapter's content and its relevance to current research trends.

With the advent of new techniques and applications, the PANI research area is set to expand further as work on sensor technology, organic solar cells, and energy storage devices improves alongside the development of hybrid materials and PANI. Moreover, the future of conducting polymers like PANI looks ever brighter, especially with the rising need for environmental friendliness and sustainability.

In conclusion, the narrative of how polyaniline or PANI evolved is a tale of the evolution of the conducting polymers field, from its inception to its increased use in many electronics and energy-related activities. Its significance lies in its ability to offer tunable electrical conductivity, high stability, and cost-effective synthesis, positioning it as a cornerstone material in the development of flexible, lightweight, and sustainable electronic devices. Continued research will further enhance its properties and expand its applications, contributing to the growth of organic electronics and green technologies.

In conclusion, this research on the electrical and electrochemical properties of polyaniline (PANI) revealed its great prospects and diversity in usage in one way or another. PANI is of great interest in electronics and energy storage because of its easy-to-synthesize, flexible, and environmentally friendly electrical conductive properties. As we further investigate, it appears that PANI's very own characteristics can be modified to a greater extent by the composites' formation with metals, metal oxides, and other nanomaterials. Shang's PANI-based composites' improved electrochemical performance makes them ideally acceptable for supercapacitors, gas sensors, rechargeable batteries, and fuel cells. Particularly interesting is the synergy of effects obtained from these specific blends that broaden the scope for the practical application of PANI in more powerful devices. Yet, dilemmas remain for PANI to be commercialized, including arguments of cost versus benefits and molecular architectures competitive enough to sustain the market. One such area that requires further work is the instability due to de-doping and structural integrity maintained under different conditions. Further, we need to strengthen how well these materials perform under electrochemical activity relative to mechanical endurance. Research in the future should focus more on developing new nanoscale materials with enhanced surface areas and better conductivities. There are possibilities of PANI or its derivatives with improved functionalities being actively used to increase the performance of supercapacitors and other batteries. There is a need to understand more about the use of PANI as a catalyst for the oxygen evolution reaction (OER) in fuel cells to know what thermoplastic elastomers are capable of doing. Additionally, PANI has potential use in other developing technologies such as nanofluids for heat transfer which could take advantage of PANI's properties to inhibit the agglomeration of nanoparticles and improve thermal conductivity. This indicates that PANI has far more uses than are currently being researched, and this alone should motivate scientists to look for novel ways of using the polymer. In essence, a conductive polymer such as PANI is likely to profoundly impact electrochemical energy storage and conversion technologies in the next-generation owing to its unique properties alongside advancement in synthesis and characterization methods. For PANI to be used commercially, there still needs to be more consistency in research and scalable manufacturing processes, but the opportunity is highly feasible.

## Funding

No funding has been received for this project.

## Competing interests

The authors declare that they have no known competing financial interests or personal relationships that could have appeared to influence the work reported in this paper.

## Ethical approval

This declaration is "not applicable."

## Availability of data and materials

Data not available/The data that has been used is confidential.

**Author details**

Alvena Shahid* and Zohra Nazir Kayani
Lahore College for Women University, Lahore, Pakistan

*Address all correspondence to: alvena.shahid@gmail.com

## References

[1] Wang H, Lin J, Shen ZX. Polyaniline (PANi) based electrode materials for energy storage and conversion. Journal of Science: Advanced Materials and Devices. 2016;**1**(3):225-255

[2] Chang C-H, Huang T-C, Peng C-W, Yeh T-C, Lu H-I, Hung W-I, et al. Novel anticorrosion coatings prepared from polyaniline/graphene composites. Carbon. 2012;**50**(14):5044-5051

[3] Meng Y, Wang K, Zhang Y, Wei Z. Hierarchical porous graphene/polyaniline composite film with superior rate performance for flexible supercapacitors. Advanced Materials. 2013;**25**(48):6985-6990

[4] Alam J, Riaz U, Ahmad S. High performance corrosion resistant polyaniline/alkyd ecofriendly coatings. Current Applied Physics. 2009;**9**(1):80-86

[5] Peng X, Zhang Y, Gong W, Zhou J, Ye D, Fan L, et al. Redox-active Janus separator based on polyaniline nanosheets and bacterial cellulose nanofibers for lithium-ion batteries. International Journal of Biological Macromolecules. 2025;**302**:140536

[6] Shahid A, Kayani ZN, Sahar M, Shahid E, Riaz S, Naseem S. Pr2CuO4/MWCNT/MXene/PANI: A novel quaternary composite with tunable properties for potential applications in optoelectronics and energy storage devices. Journal of Alloys and Compounds. 2025;**1010**:177264

[7] Muthukrishnan FP, Sundararajan PB, Murugesan A. Morphological influence of graphene and $FeSO_4$ on polyaniline and application of nanocomposite towards oxygen reduction reactions in acid medium. Hybrid Advances. 2025;**8**:100394

[8] Crowley K, Smyth MR, Killard AJ, Morrin A. Printing polyaniline for sensor applications. Chemical Papers. 2013;**67**:771-780

[9] Xiong C, Zheng C, Zhang Z, Xiong Q, Zhou Q, Li D, et al. Polyaniline@cellulose nanofibers multifunctional composite material for supercapacitors, electromagnetic interference shielding and sensing. Journal of Materiomics. 2025;**11**(1):100841

[10] Shahid A, Kayani ZN, Akram A, Anwar M, Hassan A, Sahar M. Oxygen vacancies engineering of ternary $Co_3O_4$/rGO/PANI composite for environmental and biomedical applications. Clean Technologies and Environmental Policy. 2024;**26**:1-24. DOI: 10.1007/s10098-024-03030-7

[11] MacDiarmid AG, Epstein AJ. Polyanilines: A novel class of conducting polymers. Faraday Discussions of the Chemical Society. 1989;**88**:317-332

[12] MacDiarmid AG, Epstein AJ. Conducting polymers: Science and technology. In: Frontiers of Polymers and Advanced Materials. Boston, MA: Springer US; 1994. pp. 251-261

[13] Rahman MM, Hossen MR, Alam I, Rahman MH, Faruk O, Nurbas M, et al. Synthesis of hexagonal boron nitride based PANI/h-BN and PANI-PPy/h-BN nanocomposites for efficient supercapacitors. Journal of Alloys and Compounds. 2023;**947**:169471

[14] Cao Y, Andreatta A, Heeger AJ, Smith P. Influence of chemical polymerization conditions on the

properties of polyaniline. Polymer. 1989;**30**(12):2305-2311

[15] Huang J, Kaner RB. Nanofiber formation in the chemical polymerization of aniline: A mechanistic study. Angewandte Chemie. 2004;**116**(43):5941-5945

[16] Shahid A, Kayani ZN. Recent advances in metal oxides as energy storage materials: Synthesis, characterisation, and electrochemical studies for sustainable energy storage. In: Fourth iiScience International Conference 2024: Recent Advances in Photonics and Physical Sciences. Vol. 13440. Lahore, Pakistan: SPIE; 2024. p. 1344002. DOI: 10.1117/12.3051729

[17] Feng J, Shunxian Y, Shi C, Tang X, Zhao X, Chen S, et al. Advanced cathode designs for high-energy lithium/sodium–selenium battery. Advanced Functional Materials. 2025;**35**(10):2422013. DOI: 10.1002/adfm.202422013

[18] Gospodinova N, Terlemezyan L. Conducting polymers prepared by oxidative polymerization: Polyaniline. Progress in Polymer Science. 1998;**23**(8):1443-1484

[19] Chen J, Winther-Jensen B, Pornputtkul Y, West K, Kane-Maquire L, Wallace GG. Synthesis of chiral polyaniline films via chemical vapor phase polymerization. Electrochemical and Solid-State Letters. 2005;**9**(1):C9

[20] Jun T-S, Kim C-K, Kim YS. Vapor phase polymerization of polyaniline nanotubes using Mn3O4 nanofibers as an oxidant. Materials Letters. 2014;**133**:17-19

[21] Widmann JF, Davis EJ. Photochemical initiated polymerization of single microdroplets. Colloid and Polymer Science. 1996;**274**:525-531

[22] Chen M, Zhong M, Johnson JA. Light-controlled radical polymerization: Mechanisms, methods, and applications. Chemical Reviews. 2016;**116**(17):10167-10211

[23] Yagci Y, Jockusch S, Turro NJ. Photoinitiated polymerization: Advances, challenges, and opportunities. Macromolecules. 2010;**43**(15):6245-6260

[24] Kim HS, Ahn K, Han BY, Haque A-MJ, Kim S, Kim S, et al. Conductive single enzyme nanocomposites prepared by in-situ growth of nanoscale polyaniline for high performance enzymatic bioelectrode. Biosensors and Bioelectronics. 2025;**267**:116841

[25] Shahid A, Kayani ZN, Riaz S, Naseem S. Enhanced supercapacitor potential of CeO2/MXene nanocomposites with tunable conductive polyaniline concentrations. Diamond and Related Materials. 2024b;**146**:111167

[26] Boeva ZA, Sergeyev VG. Polyaniline: Synthesis, properties, and application. Polymer Science Series C. 2014;**56**(1):144-153

[27] Bhadra J, Madi NK, Al-Thani NJ, Al-Maadeed MA. Polyaniline/polyvinyl alcohol blends: Effect of sulfonic acid dopants on microstructural, optical, thermal and electrical properties. Synthetic Metals. 2014;**191**:126-134

[28] Yakuphanoglu F, Şenkal BF. Electronic and thermoelectric properties of polyaniline organic semiconductor and electrical characterization of Al/PANI MIS diode. The Journal of Physical Chemistry C. 2007;**111**(4):1840-1846

[29] Shahid A, Kayani ZN, Nazli H, Riaz S, Naseem S. Controllable synthesis of composite electrode material to enhance the electrochemical

performance of supercapacitors. Optical Materials. 2023;**143**:114264

[30] Poudel MB, Shin M, Kim HJ. Polyaniline-silver-manganese dioxide nanorod ternary composite for asymmetric supercapacitor with remarkable electrochemical performance. International Journal of Hydrogen Energy. 2021;**46**(1):474-485

[31] Krithika S, Balavijayalakshmi J. Influence of manganese dioxide nanoparticles on MoS2/PANI Nanosheets and its energy storage applications. Journal of Inorganic and Organometallic Polymers and Materials. 2023;**33**(6):1657-1666

[32] Xie H, Ma S, He Z. Facile preparation of PANI/MoOx nanowires decorated MXene film electrodes for electrochemical supercapacitors. Electrochimica Acta. 2023;**448**:142173

[33] Wu R, Yuan H, Liu C, Lan J-L, Yang X, Lin Y-H. Flexible PANI/SWCNT thermoelectric films with ultrahigh electrical conductivity. RSC Advances. 2018;**8**(46):26011-26019

[34] Otrokhov G, Pankratov D, Shumakovich G, Khlupova M, Zeifman Y, Vasil'eva I, et al. Enzymatic synthesis of polyaniline/multi-walled carbon nanotube composite with core shell structure and its electrochemical characterization for supercapacitor application. Electrochimica Acta. 2014;**123**:151-157

[35] Ouyang L, Wang Y, Zhang P, Wang X, Yuan S. Heterostructured MWCNTs@ PANI@ TiO2 nanocomposites for enhanced adsorption of As (III) from aqueous solution: Adsorption and photocatalytic oxidation behaviors. Industrial & Engineering Chemistry Research. 2020;**59**(25):11743-11756

[36] Shao D, Hou G, Li J, Wen T, Ren X, Wang X. PANI/GO as a super adsorbent for the selective adsorption of uranium (VI). Chemical Engineering Journal. 2014;**255**:604-612

[37] Wang L, Yao Q, Bi H, Huang F, Wang Q, Chen L. PANI/graphene nanocomposite films with high thermoelectric properties by enhanced molecular ordering. Journal of Materials Chemistry A. 2015;**3**(13):7086-7092

[38] Zhu J, Chen M, Honglin Q, Zhang X, Wei H, Luo Z, et al. Interfacial polymerized polyaniline/graphite oxide nanocomposites toward electrochemical energy storage. Polymer. 2012;**53**(25):5953-5964

[39] Xiong S, Yang F, Jiang H, Ma J, Lu X. Covalently bonded polyaniline/fullerene hybrids with coral-like morphology for high-performance supercapacitor. Electrochimica Acta. 2012;**85**:235-242

[40] Palsaniya S, Nemade HB, Dasmahapatra AK. Hierarchical PANI-RGO-ZnO ternary nanocomposites for symmetric tandem supercapacitor. Journal of Physics and Chemistry of Solids. 2021;**154**:110081

[41] Radja I, Yassin MT, Ezzat AO, AbdElkader OH, Benyoucef A, Alkoudsi BD, et al. Construction of a ternary composite of mg-doped biochar/CuO, and PAni for supercapacitor applications. Journal of Energy Storage. 2025;**108**:114785

Chapter 6

# Theoretical Study of Self-Assembled Bi Nanolines on the InAs(001) Surface

*Kouloud Kourchid, Mourad Mbarki, Ramzi Alaya and Ahmed Rebey*

## Abstract

This chapter presents a theoretical investigation of the physical properties of bismuth (Bi) nanolines on the indium arsenide (InAs)(001) surface through a theoretical investigation based on first-principles calculations (Density Functional Theory (DFT)). Three structural models Miki (M), Naitoh (N), and Two Ad-Dimer (TAD) are examined to understand their atomic configurations, bond lengths, and stability. The results highlight significant differences in structural organization and emphasize the critical role of the substrate by comparing Bi nanolines on InAs(001) and silicon (Si)(001). Surface stability calculations identify the M model as the most energetically favorable configuration, suggesting its suitability for stable Bi nanoline formation on InAs(001). Electronic property analysis reveals that the M and TAD models exhibit semiconductor-like behavior, whereas the N model displays metallic characteristics. Additionally, partial electronic charge density analysis provides insights into the surface states, shedding light on the role of atomic interactions in determining electronic properties. These findings contribute to a deeper understanding of Bi nanostructures and offer guidance for their experimental realization, with potential applications in nanoscale electronics and quantum devices.

**Keywords:** bismuth nanolines, InAs(001), density functional theory (DFT), structural stability, electronic properties, surface states, semiconductor

## 1. Introduction

The field of nanotechnology has revolutionized the study of atomic-scale structures [1–3], paving the way for the development of advanced materials and devices with unprecedented functionalities [4]. At the forefront of this progress lies the exploration of semiconductor surfaces, where the self-assembly of nanostructures opens up new avenues for both fundamental research and practical applications [5]. Among these nanostructures, Bi nanolines have garnered significant attention due to their unique combination of structural precision, electronic properties, and stability [6]. This chapter focuses on the formation, structural characteristics, and electronic behavior of Bi nanolines on the (001) surface of indium arsenide (InAs), while providing a comparative analysis with similar structures on the silicon (Si)(001) surface. The interplay between the distinct properties of these substrates and the behavior of Bi atoms offers deep insights into the mechanisms governing nanostructure formation [7].

The self-assembly of Bi nanolines on the InAs(001) surface is a result of the substrate's unique polar nature and the interaction of Bi atoms with its atomic lattice [8]. InAs, a III-V semiconductor known for its high electron mobility and direct bandgap, provides an ideal platform for studying Bi-induced nanostructures [9]. The (001) surface of InAs is inherently polar, with alternating layers of indium and arsenic atoms creating an electrostatic field that influences the adsorption and organization of Bi atoms [7]. Under optimized growth conditions, including precise control of substrate temperature and Bi flux, Bi atoms organize into highly ordered, one-dimensional (1D) nanolines [10]. These structures are characterized by remarkable stability and periodicity, attributed to the interplay between the surface chemistry of InAs and the atomic properties of Bi, such as its high atomic mass and large spin-orbit coupling [11].

Structural investigations of Bi nanolines on InAs(001) employing modern methodologies, including scanning tunneling microscopy (STM) and transmission electron microscopy (TEM), reveal a periodic alignment of Bi atoms along specific crystallographic directions [12]. This alignment results in unique structural motifs that distinguish these nanolines from those on other semiconductor surfaces [13]. Complementary theoretical studies, particularly DFT calculations, have provided detailed insights into the energetics and electronic properties of these structures [14]. The electronic properties of Bi nanolines on InAs(001) are highly anisotropic, stemming from their one-dimensional nature and the quantum confinement effects induced by the substrate [15]. These properties suggest their potential for integration into quantum devices and spintronic applications, where the control of spin-orbit interactions is crucial [16].

In contrast, Bi nanolines on Si(001) exhibit a markedly different formation mechanism and set of properties [17]. Silicon, the cornerstone of modern electronics, has a covalent lattice structure and nonpolar surface, leading to distinct interactions with Bi atoms [18]. On Si(001), Bi nanolines form perpendicular to the silicon dimer rows and are stabilized by the local reconstruction of the silicon surface [19]. These nanolines demonstrate exceptional stability and periodicity, with structural widths on the order of 1 nanometer (nm) and lengths extending across the surface [20]. The absence of substrate polarity results in a fundamentally different electronic landscape for Bi nanolines on Si(001) compared to InAs(001), highlighting the critical role of substrate properties in determining the behavior of self-assembled nanostructures [21].

The choice of the InAs(001) surface for Bi nanolines, as opposed to other III-V semiconductors, is motivated by the favorable lattice matching, which leads to a more stable formation of the nanolines. InAs(001) also offers distinct electronic properties that enhance the self-assembly process of Bi nanolines, providing a promising platform for nanoscale device fabrication. Moreover, the use of InAs(001) enables the exploration of novel properties that are less achievable on other substrates. However, the experimental realization of Bi nanolines on this surface faces challenges related to precise control over growth conditions, surface reconstructions, and the interplay between Bi and the substrate, all of which influence the stability and quality of the nanolines.

The comparative study of Bi nanolines on InAs(001) and Si(001) underscores the profound influence of substrate characteristics on nanoline formation, structure, and functionality [22]. While Bi nanolines on InAs(001) benefit from the substrate's polar nature and the associated electronic anisotropy, those on Si(001) leverage the covalent bonding framework of silicon to achieve stability and uniformity [23]. These differences not only deepen our understanding of nanostructure formation but also inform the design of materials tailored for specific technological applications [24].

Bi nanolines represent a convergence of nanoscale precision and versatile functionality, offering promising applications in quantum computing, spintronics, and nanoscale electronics [25]. The interplay between their one-dimensional structure and substrate-dependent properties makes them a valuable system for exploring fundamental physical phenomena and developing next-generation devices [26]. This chapter aims to provide a comprehensive overview of Bi nanolines on InAs(001), juxtaposed with their counterparts on Si(001), to elucidate the substrate-driven mechanisms that underpin their formation and functionality. By examining these systems in detail, this work contributes to the broader understanding of semiconductor surface nanostructures and their potential impact on advanced technologies [27].

## 2. Methods of calculation

This study investigates the structural and electronic properties of self-assembled Bi nanolines on the InAs (001) surface using first-principles calculations grounded in the DFT [28]. The computational framework is built upon the pseudopotential method implemented in the Quantum Espresso code [29], which employs pseudopotentials with conserved norms. The electronic wave function is expanded in a plane-wave basis, which is particularly well suited for simulating the physical and chemical properties of materials in a computationally efficient manner. This method provides accurate results for the interactions of atoms, electronic structure, and the formation of nanostructures.

For the treatment of exchange and correlation interactions, we use the local density approximation (LDA) [30], parameterized by Perdew-Zunger [31], which has been widely used for solid-state systems. The LDA is known for its accuracy in describing the electronic structure of materials with relatively simple bonding, making it an appropriate choice for the current study. However, it is well known that LDA tends to underestimate the bandgap of semiconductors due to its inherent approximations in treating exchange-correlation interactions. Despite this limitation, LDA remains a reliable choice for surface energy calculations and structural predictions, making it well suited for our investigation of Bi nanolines on the InAs (001) surface. The atomic configurations of the elements involved in the system are as follows:

- *Indium (In):* [Kr] $4d^{10}$ $5s^2$ $5p^1$
- *Arsenic (As):* [Ar] $3d^{10}$ $4s^2$ $4p^3$
- *Bismuth (Bi):* [Ar] $5d^{10}$ $6s^2$ $6p^3$

To simulate the self-assembly of Bi nanolines on the InAs (001) surface, we employed the supercell method, a robust and widely used technique in surface science and nanomaterial simulations. The supercell method involves constructing a periodic unit cell that mimics the surface system, where the slab is composed of several atomic layers with a vacuum region separating periodic images of the surface. This approach is ideal for studying surface phenomena and nanostructures, as it allows for the simulation of realistic surface structures and properties under periodic boundary conditions.

The InAs (001) surface was modeled using a periodic slab approach, where each slab is oriented along the [001] direction. The supercell consists of nine atomic layers of InAs, with a vacuum thickness of 15 Å to avoid interactions between periodic images of the surface [32]. The choice of 15 Å for the vacuum region ensures that the slabs are sufficiently separated to minimize any spurious interaction between the slabs, while still maintaining the periodicity needed to simulate bulk-like behavior in the system. This method accurately simulates the surface properties while excluding any bulk contributions from the subsurface regions.

In terms of Brillouin zone integration, a (2 × 4 × 1) k-point grid was used to sample the zone, following the Monkhorst-Pack method [33]. The Monkhorst-Pack technique provides an efficient and systematic way to sample the Brillouin zone while maintaining the precision of the results. The selection of a (2 × 4 × 1) k-point grid strikes a balance between computational efficiency and accuracy, particularly for slab calculations, where fewer k-points are often sufficient due to the relatively simple electronic structure along the surface direction.

The kinetic energy cutoff of 40 Ry (Rydberg) is used to truncate the plane-wave basis set, which is standard for achieving convergence in typical materials calculations. This cutoff ensures that the plane-wave expansion of the wave functions is adequately represented, and that the total energy is well converged with respect to the cutoff. Structural optimization was conducted on the upper four atomic levels of the slab from the In-terminated side, as these are most sensitive to surface effects, such as relaxation and reconstruction. The lower layers were fixed during optimization to maintain the bulk-like properties of the material and to minimize computational cost. This optimization process allows for the relaxation of atomic positions, which is critical for capturing the true surface energy and morphology of the Bi nanolines.

The supercell approach, combined with the careful selection of computational parameters, provides an accurate and efficient way to model the self-assembled Bi nanolines on the InAs (001) surface. The ability to simulate and optimize the atomic structure of these nanolines, while maintaining periodic boundary conditions and accurately calculating electronic properties, makes this methodology highly suitable for studying nanostructures at the atomic scale. This method is anticipated to produce significant insights into the creation mechanisms and electrical properties of Bi nanolines, establishing a basis for forthcoming experimental studies and technological applications.

## 3. Structural properties

Miki et al. [34] developed the Miki model by meticulous examination of the STM images, characterizing Bi nanolines on the Si(001) surface as consisting of two parallel Bi dimers, with a vacant dimer row situated between them. Conversely, the N model, introduced by Naitoh et al. [34–36], presents an alternative configuration, in which Bi nanolines form through the substitution of a set of four Si dimers with two parallel, adjacent Bi dimers, accompanied by an absent dimer row next to each Bi dimer.

To model the generation of Bi nanolines on the InAs(001)-(2 × 6) reconstructed surface, we especially examined three models in our investigation: M, N, and TAD models [37–40]. **Figure 1** vividly illustrates the self-assembled Bi nanolines on the InAs(001) surface as conceptualized by these three models, providing a comparative visual representation of their structural characteristics (**Figure 2**).

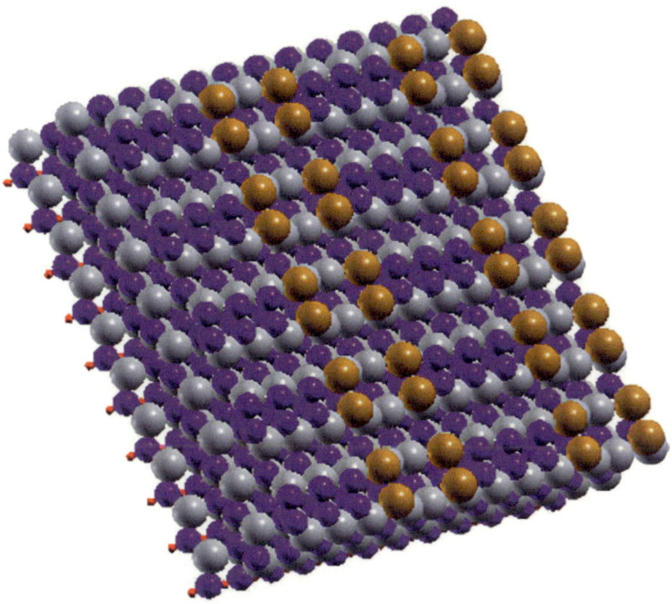

**Figure 1.**
*Bi nanolines deposited on the InAs(001)-(2 × 6) surface.*

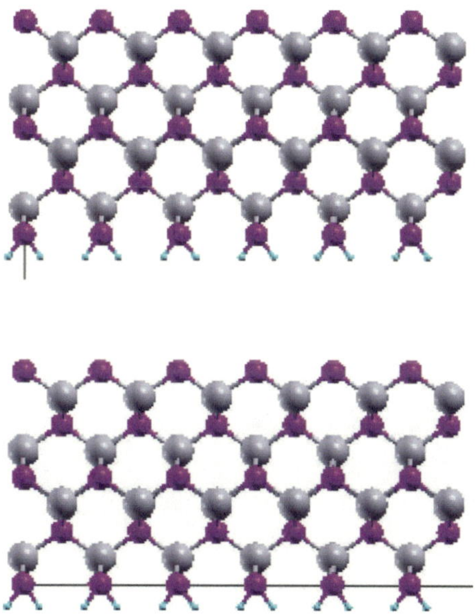

**Figure 2.**
*Supercell model of InAs (001).*

**Figure 3** compares the atomic structures of Bi nanolines on the InAs(001)-(2 × 6) surface across three models: M, N, and TAD. In the M model (See **Figure 3a**), Bi dimers are aligned along [−110], perpendicular to In dimers, with a separation of one In dimer. The N model (See **Figure 3b**) includes Bi nanolines that are perpendicular to [110], the removal of an As dimer in the top layer, and the production of In–In dimers

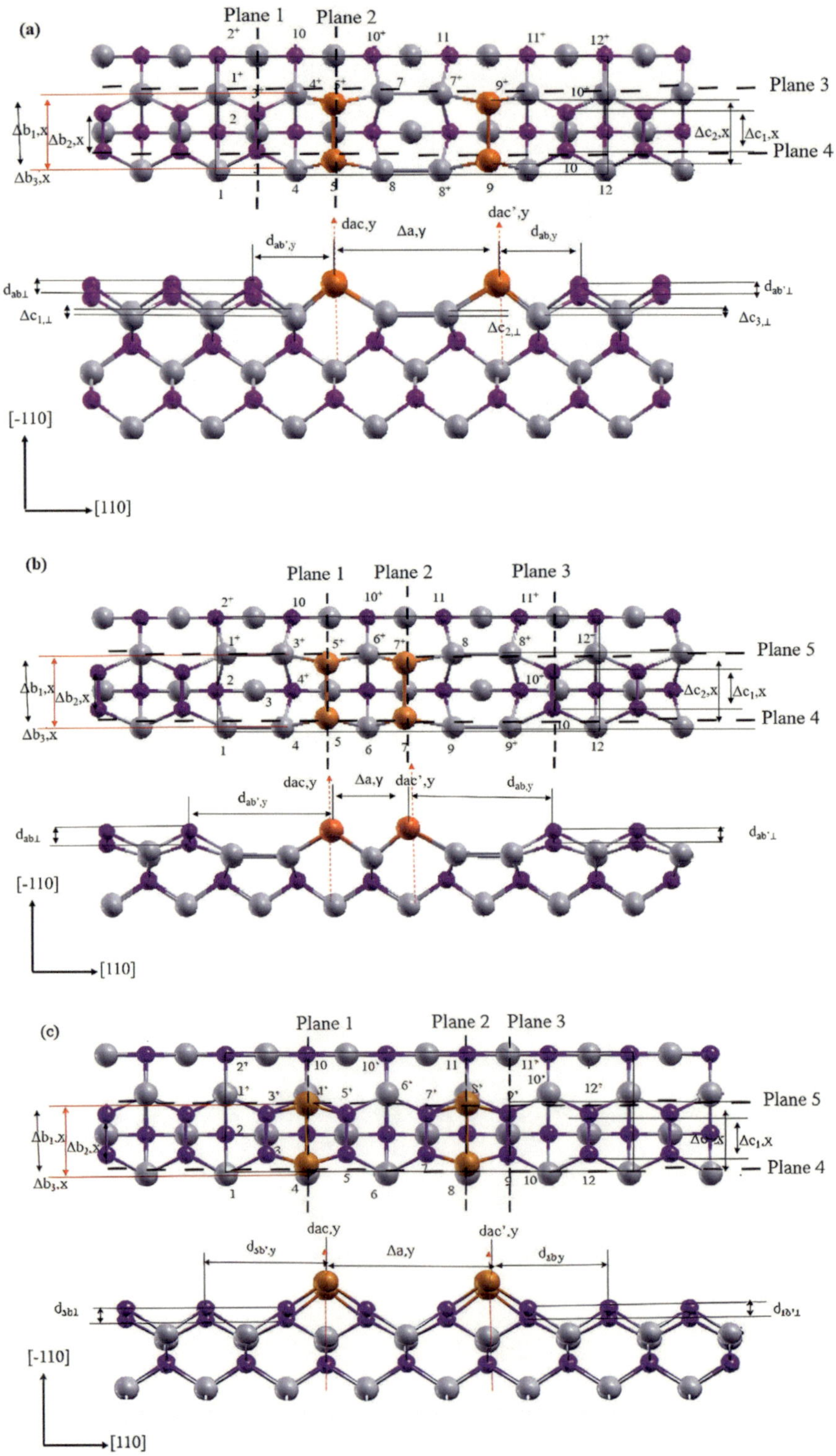


**Figure 3.**
*Top and side views depicting the Bi nanolines on the InAs(001) (2x6) surfaces, illustrating the M, N, and TAD models.*

in the second layer. Both models have a 3.16 Å bond length between Bi(5) and Bi(5+), which is compatible with the M model.

The TAD model (**Figure 3c**) shows a slightly longer Bi(8)–Bi(8+) bond of 3.19 Å and unique interactions, such as In(1)–As(2) at 2.78 Å and In(1+)–As(2) at 2.57 Å. Its As dimer buckling is more pronounced (0.76 Å) than in the M (0.62 Å) and N (0.67 Å) models. The Bi nanoline spacing on InAs(001) (8.44 Å) is larger than on Si(001) (6.30 Å), indicating a more relaxed configuration.

Distinctive features like In-In dimers in the N model and additional interactions in the TAD model highlight substrate-specific behaviors. These findings align with prior studies [41, 42], emphasizing how the substrate influences the structural and electronic properties of Bi nanolines.

**Figure 3b** depicts the atomic structure of the N model, which differs slightly from that of the M model depicted in **Figure 4a**. According to the N model, Bi nanolines are formed by aligning two parallel Bi dimers in the [−110] direction. These Bi dimers replace As dimers, resulting in a structure without a dimer row next to each Bi dimer. The bond length of Bi dimers in the N model is 3.16 Å, which is consistent with the value given for the M model [34–48]. In comparison, the Bi dimers in the third model, the TAD model, are slightly longer, with an increase of around 0.03 Å. To assess the stability of Bi nanolines on the pure InAs(001) surface, the total energies of the three models were thoroughly compared. This method sought to determine the energetically optimal structure for the Bi nanolines by evaluating the total energy differences among the three models examined in our study (M, N, and TAD models), as described in McLean et al. [35].

$$\Delta E = E(M) - E(N) - 2\mu_{\mathrm{InAs}}^{\mathrm{Bulk}} \quad (1)$$

The total energies of the M and N models are denoted by E(M) and E(N), respectively, while $\mu_{InAs}^{Bulk}$ represents the chemical potential of bulk InAs. For consistency, $\mu_{InAs}^{Bulk}$ was determined using the same method applied in the surface (Bi-line) energy calculations. According to our analysis, the M model is 0.21 eV per dimer more energetically favorable than the N model, which translates to $\Delta E$ = -0.21. Likewise, the M model is 0.12 eV per dimer more stable than the TAD model, or $\Delta E$ = -0.12 eV/dimer, which is consistent with the findings of Miwa et al. [49].

## 4. Electronic properties

**Figure 5** depicts the electronic band structure of Bi nanolines self-assembled on the InAs(001) surface, plotted within the first Brillouin zone along high-symmetry points (Γ-J-K-J′-Γ). The bulk band structure of InAs is shown as a shaded region, allowing for a direct comparison with the electronic states associated with the Bi nanolines. The study investigates three configurations: M, N, and TAD. Both the M and N models exhibit five distinct nondispersive surface states ($V_1$–$V_5$) that are occupied, while the TAD model shows only three occupied surface states ($V_1$–$V_3$). For clarity, the unoccupied states are labeled as $C_1$ and $C_2$.

**Figure 5a** displays the band structure for the M model, where the occupied states V1–V4 are situated below the Fermi level. Notably, V1 is positioned 0.10 eV above the bulk valence band maximum, indicating a semiconductor-like behavior. In **Figure 5b**,

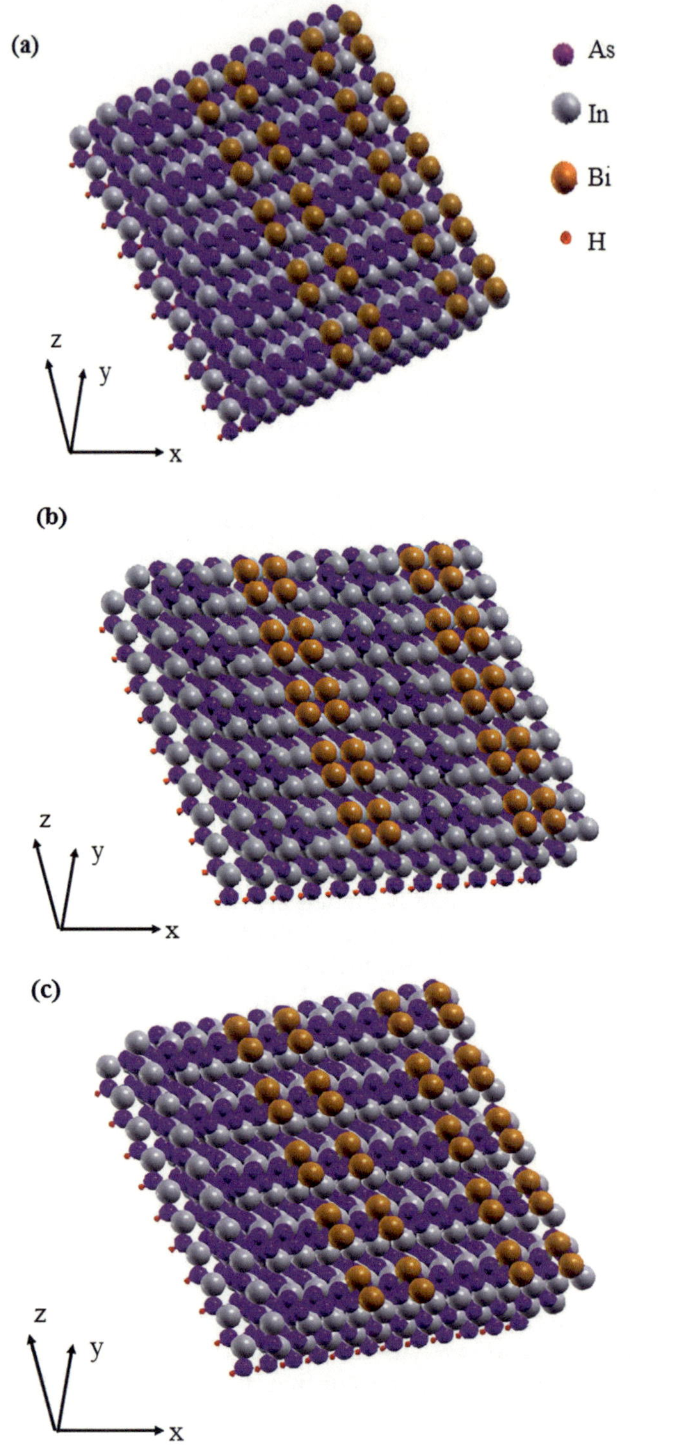


**Figure 4.**
*Bi nanoline deposited on InAs (001) (2x6) surfaces. (a) M model, (b) N model, and (c) TAD model.*

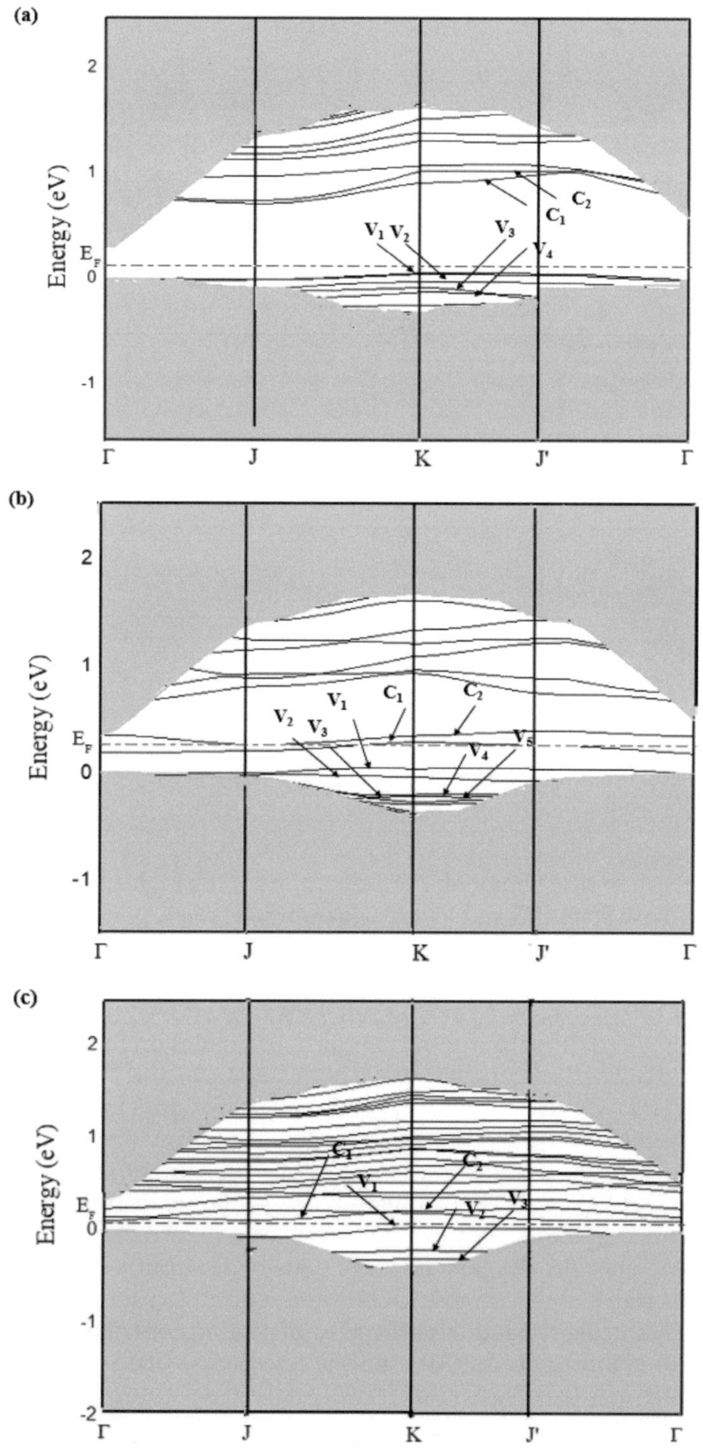


**Figure 5.**
*Electronic band structures of the Bi nanoline on the InAs (001) -(2x6) surface for (a) the M model, (b) the N model, and (c) the TAD model.*

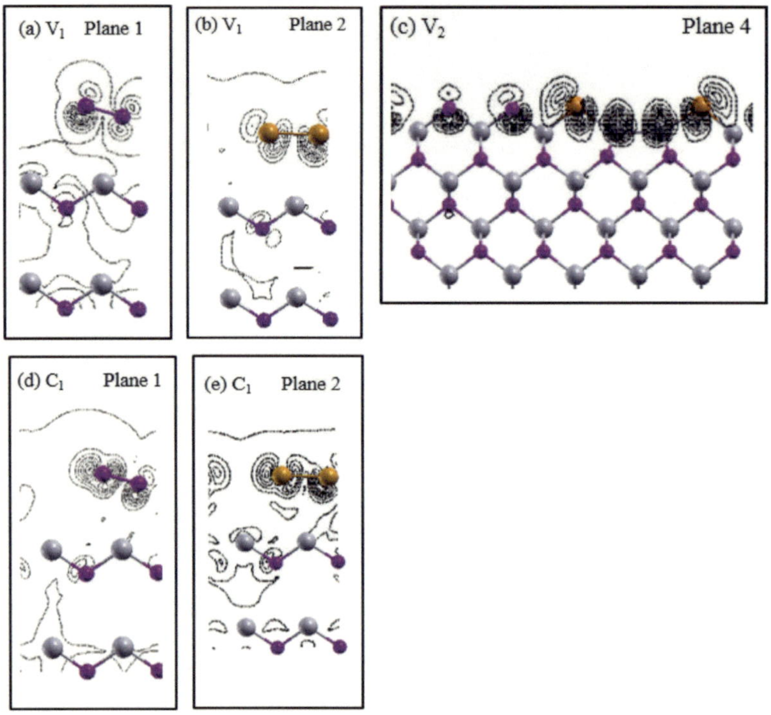


**Figure 6.**
*Partial electronic charge density plots of M model plots at the K point for: (a), (b) occupied state $V_1$, and (c) occupied state $V_2$, and (d), (e) of unoccupied state $C_1$.*

the band structure for the N model is presented, with the highest surface state, V1, located at 0.034 eV. States V3 and V4 are degenerate at the K point and lie approximately 0.2 eV below the valence band maximum. Additionally, the unoccupied surface state $C_1$ intersects the Fermi level, indicating metallic behavior. This suggests that Bi nanolines on InAs(001) could exhibit inherent conductivity. A study by Nafday et al. [14] investigated three configurations of Bi nanolines on InAs(001)—(2 × 1), β(4 × 2), and ξ(4 × 2)—and concluded that both Bi/(2 × 1) and Bi/β(4 × 2) exhibit metallic properties, whereas Bi/ξ(4 × 2) retains its semiconductor characteristics.

In the TAD model, all occupied states lie below the Fermi level, and no conduction states intersect it, confirming its semiconductor nature (refer to **Figure 3c**). However, it is essential to highlight that DFT using the LDA is known to underestimate the bandgaps of semiconductors.

**Figure 6(a-e)** presents the partial charge density distributions for both occupied and unoccupied states in the M model. The highest occupied surface state, $V_1$, is mainly driven by the hybridization of the $\pi^*$ antibonding contributions from the px orbitals of the Bi-stabilized InAs(001) surface (see **Figure 6b**) and the pz orbitals of the top-layer As dimer (see **Figure 6a**). The $V_2$ state arises from the antibonding $\pi^*$ contributions of the px orbitals from the Bi dimer in the top layer, with additional influence from the In atom in the second layer. The conduction state $C_1$ is predominantly attributed to the bonding $\pi^*$ contributions of the px orbitals from the top-layer Bi atom, with substantial involvement from the arsenic atoms. These findings align with recent studies [32, 50, 51], reinforcing the reliability of our analysis and deepening our understanding of the electronic structure and bonding interactions.

**Figure 7** presents the partial charge density of the N model, showing the origins of its electronic states. The highest occupied state, $V_1$, primarily originates from the $\pi^*$ antibonding interactions of the pz orbitals in the top-layer As atoms (See **Figure 7a**). Meanwhile, $V_2$ is linked to the $\sigma$ orbital formed by the Bi–Bi bond in the top layer (See **Figure 7b**). The lowest unoccupied state, $C_1$, is largely derived from the $\pi^*$ antibonding contributions of the $p_x$ orbitals associated with the Bi–Bi dimers in the top layer, with a smaller contribution coming from the $p_z$ orbitals of the As atoms (See **Figure 7c** and **d**).

**Figure 8** illustrates the partial charge density for the TAD model, highlighting the origins of its electronic states. The highest valence state, $V_1$, is mainly attributed to the $\pi^*$ antibonding contributions of the $p_z$ orbitals from the Bi dimer in the top layer (**Figure 8a** and **b**). The surface state $V_2$ results from the antibonding interactions between the $p_z$ orbitals of the As atoms in the top layer (**Figure 8c**). Meanwhile, the lowest unoccupied state, $C_1$, originates from the $\pi^*$ antibonding contributions of the $p_x$ orbitals from the Bi atoms in the top layer (**Figure 8d**).

Our calculations reveal that both the M and TAD models exhibit semiconductor behavior, making them promising candidates for serving as templates in the growth

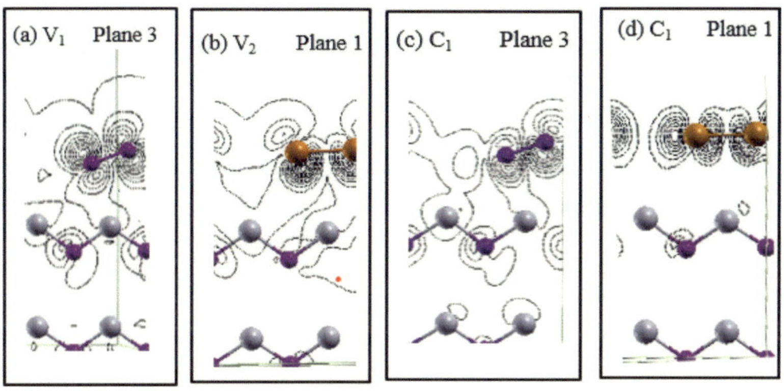


**Figure 7.**
*The partial electronic charge density plots for the N model at the K point are presented as follows: (a) the occupied state V1, (b) the occupied state V2, and (c) and (d) the unoccupied state C1.*

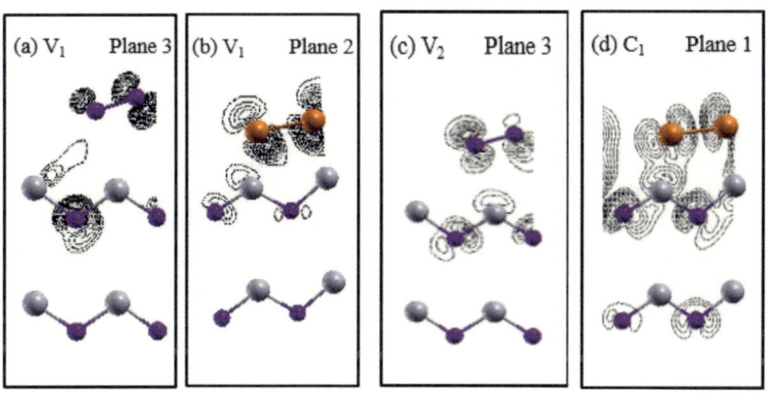


**Figure 8.**
*Partial electronic charge density plots of TAD model at the K point for: (a) and (b) occupied state $V_1$, and (c) occupied state $V_2$, and (d) of unoccupied state $C_1$.*

of one-dimensional structures with customizable electrical and optical properties. On the other hand, the N model displays metallic characteristics due to energy states near the Fermi level (**Figure 3b**), suggesting that Bi nanolines on InAs(001) could serve as conductive nanowires in prototype electronic devices.

Our findings are noteworthy for their potential applications. The calculations indicate that the M and TAD models possess semiconductor properties, which could be highly beneficial for developing optical and electro-optical devices. On the other hand, the N model exhibits metallic behavior, suggesting that Bi nanolines in this configuration can conduct electricity independently at low bias voltage. This characteristic positions them as promising candidates for nanowires in prototype electronic devices.

## 5. Conclusion

This study employs DFT calculations to investigate the structural and electronic properties of Bi nanolines on the InAs (001) surface, focusing on three structural models: Miki, Naitoh, and Two Ad-Dimer (TAD). Key physical parameters, including atomic arrangements, bond lengths, and surface organization, were analyzed to reveal the underlying mechanisms governing nanoline formation. A comparative analysis with Bi nanolines on Si (001) highlights the significant role of the substrate in stabilizing and shaping these nanostructures. The findings indicate that the Miki model is the most stable configuration, making it the most promising for potential applications. Electronic property analysis further demonstrates that the Miki and TAD models exhibit semiconductor characteristics, while the N model displays metallic behavior, suggesting its suitability as a conductive nanowire.

Beyond the structural and electronic characterization, this study highlights the challenges associated with the experimental realization of Bi nanolines. The precise control of growth parameters, such as substrate temperature and Bi flux, is critical for achieving well-ordered, defect-free nanolines. Additionally, substrate-induced strain and atomic-scale reconstructions must be carefully managed to ensure reproducibility in fabrication. Addressing these factors is essential for translating theoretical predictions into experimentally viable nanostructures.

Future research could extend beyond the LDA approach by incorporating advanced computational techniques such as GW approximations or hybrid functionals to enhance the accuracy of bandgap predictions and electronic structure modeling. These improvements could provide a deeper understanding of the electronic properties of Bi nanolines and further support their integration into functional nanoscale devices.

Bi nanolines hold significant potential for applications in spintronic devices, quantum wires, and nanoscale electronics. Their strong spin-orbit coupling and one-dimensional nature make them particularly promising for next-generation quantum technologies. Exploring their integration into practical device architectures remains a key challenge, but overcoming these hurdles could pave the way for novel advancements in semiconductor and nanoelectronic applications.

## Author details

Kouloud Kourchid[1], Mourad Mbarki[1,2*], Ramzi Alaya[1,3] and Ahmed Rebey[4]

1 Laboratory for Physics of Materials and Nanomaterials Applied to the Environment (LaPhyMNE), Faculty of Science of Gabes, University of Gabes, Gabes, Tunisia

2 Physics Department, Faculty of Science of Jeddah, University of Jeddah, Jeddah, Saudi Arabia

3 Science Department, Rustaq College of Education, University of Technology and Applied, Sciences, Sultanate of Oman

4 College of Science, Qassim University, Buraydah, Qassim, Saudi Arabia

*Address all correspondence to: mourad.mbarki@issatgb.rnu.tn

## References


[1] Miki K, Bowler DR, Owen GHJ, Briggs DAG, Sakamoto K. Atomically perfect bismuth lines on Si (001). Physical Review B. 1999;**59**(22):14868-14871. DOI: 10.1103/PhysRevB.59.14868

[2] Naitoh M, Shimaya H, Nishigaki S, Oishi N, Shoji F. Scanning tunneling microscopy observation of bismuth growth on Si(100) surfaces. Surface Science. 1997;**377**(C):899. DOI: 10.1016/S0039-6028(96)01518-X

[3] Owen GHJ, Miki K, Koh H, Yeom HW, Bowler DR. Stress relief as the driving force for self-assembled Bi nanolines. Physical Review Letters. 2002;**88**(22):226104. DOI: 10.1103/PhysRevLett.88.226104

[4] Zou P, Hong X, Chu X, Li Y, Liu Y. Multifunctional $Fe_3O_4$/ZnO nanocomposites with magnetic and optical properties. Journal of Nanoscience and Nanotechnology. 2010;**10**(3):1992-1997. DOI: 10.1166/jnn.2010.2098

[5] López LM, Cruz-Hernández E, Martínez-Velis I, Rojas-Ramírez J, Ramírez-López M, Pulzara-Mora A, et al. Self-assembly of semiconductor nanostructure. Respuestas. 2007;**12**(2):17-26

[6] Chou C, Wu B-X, Lin H-H. Structural properties of Bi thin film grown on Si (111) by quasi-van der Waals epitaxy. Scientific Reports. 2022;**12**(1):2764. DOI: 10.1038/s41598-022-06472-5

[7] Gurlu Ö, Adam OAO, Zandvliet HJW, Poelsema B. Self-organized, one-dimensional Pt nanowires on Ge (001). Applied Physics Letters. 2003;**83**:4610. DOI: 10.1063/1.1630383

[8] Tixier S, Adamcyk M, Tiedjec T, Francoeur S, Mascarenhas A, Peng W, et al. Molecular beam epitaxiy growth of GaAsBi. Applied Physics Letters. 2003;**82**:2245. DOI: 10.1063/1.1565499

[9] Miki K, Bowler DR, Owen GHJ, Briggs GAD, Sakamoto K. Structure of atomically perfect lines of bismuth in the Si (001) surface. Physical Review B. 2000;**62**(11):7237-7240. DOI: 10.1103/PhysRevB.62.7237

[10] Masnadi-Shirazi M, Lewis RB, Bahrami-Yekta V, Tiedje T, Chicoine M, Servati P. Bandgap and optical absorption edge of $GaAs_{1-x}Bi_x$ alloys with $0 < x < 17.8\%$. Journal of Applied Physics. 2014;**116**:223506. DOI: 10.1063/1.4904081

[11] Schmidt WG. III–V compound semiconductor (001) surfaces. Applied Physics A. 2002;**75**(1):89-99. DOI: 10.1007/s003390101058

[12] Ahola-Tuomi M, Punkkinen MPJ, Laukkanen P, Kuzmin M, Lang J, Schulte K, et al. Properties of self-assembled Bi nanolines on InAs(100) studied by core-level and valence-band photoemission, and first-principles calculations. Physical Review B. 2011;**83**(24):24540. DOI: 10.1103/PhysRevB.83.245401

[13] Owen JHG, Miki K, Bowler DR. Self-assembled nanowires on semiconductor surfaces. Journal of Materials Science. 2006;**41**:4568-4903. DOI: 10.1007/s10853-006-0246-x

[14] Nafday D, Richter C, Heckmann O, Wang W, Mariot JM, Djukic U, et al. Electronic structure of Bi nanolines on InAs(100). Applied Surface Science. 2023;**611**:155436. DOI: 10.1016/j.apsusc.2022.155436

[15] Leiva EP, Sanchez CG, Velez P, Schmickler W. Theory of electrochemical

monatomic nanowires. Physical Review B. 2006;**74**:035422. DOI: 10.1103/PhysRevB.74.035422

[16] Gumhalter B, Siber A, Buljan H, Fauster T. Non adiabatic dynamics of electron scattering from adsorbates in surface bands. Physical Review B. 2008;**78**:155410. DOI: 10.1103/PhysRevB.78.155410

[17] Kourchid K, Alaya R, Mbarki M, Althaqafi Y, Rebey A. Exploring the structural and electronic properties of self-assembled bismuth nanolines on InAs(001) surface through first-principles calculations. Materials Science in Semiconductor Processing. 2024;**158**:108603. DOI: 10.1016/j.mssp.2024.108603

[18] Kirkham CJ, Longobardi M, Köster SA, Renner C, Bowler DR. Subatomic electronic feature from dynamic motion of Si dimer defects in Bi nanolines on Si(001). Physical Review B. 2017;**96**(7):075304. DOI: 10.1103/PhysRevB.96.075304

[19] Murata K, Yagi S, Kanazawa T, Tsubomatsu S, Kirkham C, Nittoh K, et al. Activation of two dopants, Bi and Er in δ-doped layer in Si crystal. Nano Futures. 2022;**5**(4):045005. DOI: 10.1088/2399-1984/ac421d

[20] Bowler DR, Owen JHG. Structure of Bi nanolines: Using tight-binding to search parameter space. Condensed Matter. 2002;**14**:6761. DOI: 10.48550/arXiv.cond-mat/0202530

[21] Song B, Zhi G, Hua C, Wu M, Dou W, Gao W, et al. One-dimensional topological phase and tunable soliton states in atomic nanolines on Si(001) surface. npj Quantum Materials. 2024;**9**:24. DOI: 10.1038/s41535-024-00637-3

[22] Owen JHG, Miki K, Bowler DR. Interaction between electronic structure and strain in Bi nanolines on Si(001). Surface Science. 2003;**527**(1-3):375-382. DOI: 10.1016/S0039-6028(03)00029-3

[23] Owen JHG, Miki K, Bowler DR. Self-assembled nanowires on semiconductor surfaces. Journal of Materials Science. 2005;**41**(14). DOI: 10.1007/s10853-006-0246-x

[24] Huang H, Liu J, Duan W. Nontrivial topology in bismuth-based III–V compounds. Physical Review B. 2014;**90**(19):195105. DOI: 10.1103/PhysRevB.90.195105

[25] Miwa RH, Srivastava GP. Self-organized Bi lines on the Si(001) surface: A theoretical study. Physical Review B. 2002;**66**:235317. DOI: 10.1103/PhysRevB.66.235317

[26] Longobardi M, Kirkham CJ, Villarreal R, Köster SA, Bowler DR, Renner C. Electronic coupling between Bi nanolines and the Si(001) substrate: An experimental and theoretical study. Physical Review B. 2017;**96**(23):235421. DOI: 10.1103/PhysRevB.96.235421

[27] Sheverdyaeva PM, Bihlmayer G, Modesti S, Feyer V, Jugovac M, Zamborlini G, et al. Giant Rashba-splitting of one-dimensional metallic states in Bi dimer lines on InAs(100). Condensed Matter. arXiv:2404.13276. 2024. DOI: 10.48550/arXiv.2404.13276

[28] Mourik TV, Bühl M, Gaigeot MP. Density functional theory across chemistry, physics and biology. Philosophical Transactions of the Royal Society A. 2014;**372**:20120488. DOI: 10.1098/rsta.2012.0488

[29] Giannozzi P, Baseggio O, Bonfà P, Brunato D, Car R, Carnimeo I, et al. Quantum ESPRESSO toward the exascale. The Journal of Chemical

Physics. 2020;**152**:154105. DOI: 10.1063/5.0005082

[30] Giese TJ, York DM. Density-functional expansion methods: Evaluation of LDA, GGA, and meta-GGA functionals and different integral approximations. Journal of Chemical Physics. 2010;**133**:244107. DOI: 10.1063/1.3515479

[31] Perdew JP, Zunger A. Self-interaction correction to density-functional approximations for many-electron systems. Physical Review B. 1981;**23**:5048. DOI: 10.1103/PhysRevB.23.5048

[32] Bhat SS, Bhattacharjee S, Choi J-H, Lee S-C. Influence of surface structure and strain on indium segregation in InGaAs: A first-principles investigation. Condensed Matter. arXiv:2010.14106. 2020. DOI: 10.48550/arXiv.2010.14106

[33] Monkhorst HJ, Pack JD. Special points for Brillouin-zone integrations. Physical Review B. 1976;**13**:5188. DOI: 10.1103/PhysRevB.13.5188

[34] Naitoh M, Nishigaki S, Oishi N, Shoji F. Structure of Bi-dimer linear chains on a Si(100) surface: A scanning tunneling microscopy study. Japanese Journal of Applied Physics. 2000;**39**:2793. DOI: 10.1143/JJAP.39.2793

[35] McLean AB, Hill IG, Lipton-Duffin JA, MacLeod JM, Miwa RH, Srivastava GP. Nanolines on silicon surfaces. International Journal of Nanotechnology. 2008;**5**:1018-1057. DOI: 10.1504/IJNT.2008.01983

[36] Naitoh M, Shimiya H, Nishigaki S, Oishi N, Shoji F. Bismuth-induced surface structure of Si(100) studied by scanning tunneling microscopy. Applied Surface Science. 1999;**142**:38-42. DOI: 10.1016/S0169-4332(98)00635-7

[37] Owen JHG, Miki K, Bowler DR. Self-assembled nanowires on semiconductor surfaces. Journal of Materials Science. 2006;**41**:4568-4603. DOI: 10.1007/s10853-006-0246-x

[38] Bowler DR, Owen JHG. Structure of Bi nanolines: Using tight binding to search parameter space. Journal of Physics: Condensed Matter. 2002;**14**:6761-6769. DOI: 10.1088/0953-8984/14/26/314

[39] Naitoh M, Takei M, Nishigaki S, Oishi N, Shoji F. Structure of Bi-dimer linear chains on a Si(100) surface: A scanning tunneling microscopy study. Japanese Journal of Applied Physics. 2000;**39**:2793-2794. DOI: 10.1143/JJAP.39.2793

[40] Miwa RH, MacLeod JM, Srivastava GP, McLean AB. The equilibrium geometry and electronic structure of Bi nanolines on clean and hydrogenated Si(001) surfaces. Nanotechnology. 2005;**16**(10):2427-2435. DOI: 10.1088/0957-4484/16/10/073

[41] Tuomi MA, Punkkienen MPJ, Laukkanen P, Kuzmin M, Lang J, Schulte K, et al. Properties of self-assembled Bi nanolines on InAs(100) studied by core-level and valence-band photoemission, and first-principles calculations. Physical Review B. 2011;**83**:245401. DOI: 10.1103/PhysRevB.83.245401

[42] Zdetsis AD, Koukaras EN, Zdetsis DA. *Ab initio* study of the silicon–bismuth interface: Bismuth nanolines and bismuth-covered silicon nanoparticles. Microelectronic Engineering. 2012;**90**:99-103. DOI: 10.1016/j.mee.2011.05.031

[43] Segovia P, Purdie D, Hengsberger M, Baer Y. Observation of spin and charge collective modes in one-dimensional

metallic chains. Nature. 1999;**402**:504-507. DOI: 10.1038/990052

[44] Tuomi MA, Laukkanen P, Punkkinen M, Perälä R, Väyrynen I, Kuzmin M, et al. Formation of an ordered pattern of Bi nanolines on InAs(100) by self-assembly. Applied Physics Letters. 2008;**92**:011926. DOI: 10.1063/1.2831691

[45] AlZahrani AZ, Srivastava GP. Density-functional calculations for self-assembled Bi-nanolines on the InAs(100) surface. Journal of Applied Physics. 2009;**106**:053713. DOI: 10.1063/1.3213373

[46] Owen JHG, Miki K, Bowler DR. Interaction between electronic structure and strain in Bi nanolines on Si(0 0 1). Surface Science. 2003;**527**(1-3):L177-L183. DOI: 10.1016/S0039-6028(03)00029-3

[47] MacLeod JM, Miwa RH, Srivastava GP, McLean AB. The electronic origin of contrast reversal in bias-dependent STM images of nanolines. Surface Science. 2005;**576**:116-122. DOI: 10.1016/j.susc.2004.11.045

[48] Tajiri H, Yashiro W, Sakata O, Sakamoto K, Miki K. Transmission X-ray diffraction from bismuth lines embedded in silicon. Transactions of the Materials Research Society of Japan. 2008;**33**(3):619-622. DOI: 10.14723/tmrsj.33.619

[49] Miwa RH, MacLeod JM, Srivastava GP, McLean AB. The equilibrium geometry and electronic structure of Bi nanolines on clean and hydrogenated Si(001) surfaces. Nanotechnology. 2005;**16**:2427. DOI: 10.1088/0957-4484/16/10/073

[50] Kourchid K, Alaya R, Rebey A, Mbarki M. Adsorption of Bi adatom on InAs (001) – $\beta_2$ (2 × 4) and $\alpha_2$ (2 × 4) surface: A first principles study. Materials Science in Semiconductor Processing. 2020;**107**:104856. DOI: 10.1016/j.mssp.2019.104856

[51] Kourchid K, Mbarki M, Alaya R, Rebey A. Theoretical study of adsorption of Bi on cation-rich InAs/(0 0 1) and InP (0 0 1)- ζ (4 × 2) reconstructed surfaces. Materials Science and Engineering B. 2024;**302**:117231. DOI: 10.1016/j.mseb.2024.117231